Design and Architectures for Digital Signal Processing

Design and Architectures for Digital Signal Processing

Editor

Edvin Falk

scitus academics

Design and Architectures for Digital Signal Processing
Edited by **Edvin Falk**

ISBN: 978-1-68117-290-3
Library of Congress Control Number: 2016937466

SCITUS Academics LLC,
www.scitusacademics.com
Box No. 4766, 616 Corporate Way,
Suite 2, Valley Cottage,
NY 10989

Printed in United States of America on Acid Free Paper

Preface

The design and architecture of digital signal processing (DSP) refers to a variety of techniques to improve the accuracy and reliability of digital communication. The theory of DSP is quite complex. In essence, DSP works by clarifying or standardizing digital signals. The ADSP circuit can distinguish between orderly signals and chaotic noise inherently. The digital signals and analog signals processing are subfields of signal processing. DSP applications include audio and voice signal processing, sonar and radar signal processing, sensor array processing, spectral estimation, digital image processing, signal communication processing, system control, biomedical signal processing, seismic data processing, etc. DSP algorithms have been used on standard computers and on specialized processors called digital signal processors for a long time, and on purpose-built hardware such as an integrated application system. Additional technologies are currently used to process digital signals, including more powerful general purpose microprocessors, programmable field gate arrays, digital signal controllers (mainly for industrial applications such as motor control) and stream processors. Linear or non-linear operations can be used in the processing of digital signal. Nonlinear signal processing is closely linked to the identification of nonlinear systems and can be performed in time, frequency and time-space domains. The increased use of computers led to a greater use and need for digital signal processing. Digital signal processing addresses a wide range of applications in which high-performance systems fulfilling stringent requirements and performance constraints are receiving increasing attention in industry and academia.

The purpose of this book is to provide students, researchers, engineers and industry with a guide to the latest innovations in the development and implementation of DSP systems for specific circuits and programmable appliances.

Table of Contents

CHAPTER 1

Optical Signal Processing: Data Exchange

Jian Wang[1] and Alan E. Willner[2]

1 Wuhan National Laboratory for Optoelectronics, College of Optoelectronic Science and Engineering, Huazhong University of Science and Technology, Wuhan, Hubei, China

2 Department of Electrical Engineering, University of Southern California, Los Angeles, California, USA

1. INTRODUCTION

Optical signal processing is considered to be an attractive technique to enable fast signal manipulation in the optical domain which can avoid cumbersome optical-electrical-optical (OEO) conversions [1]. Driven by the rapid increase of traffic rates, network capacity and complexity, advanced optical networks raise the significance of data traffic grooming and require different optical signal processing functions at network nodes to achieve enhanced network efficiency and flexibility. Typical optical signal processing operations include wavelength conversion, logic gate, format conversion, delay for buffer, regeneration, add/drop, (de)multiplexing, multicasting, etc [2-14]. One may note that most of these functions work in a similar fashion of unidirectional information transfer. For example, wavelength conversion copies information from one wavelength and transfers it onto another wavelength [2]. To achieve superior network performance, bidirectional information swapping, named data exchange, would be expected to provide enhanced flexibility of optical signal processing compared to unidirectional information transfer [15].

Generally speaking, as an important concept for efficiently utilizing network resources and improving network performance, data exchange refers to the information swapping between different wavelengths/ time-slots/polarizations or other degrees of freedom. In the wavelength domain (e.g., wavelength-division multiplexed (WDM) network), data exchange, which is also known as wavelength exchange or wavelength interchange, would require the swapping of data from one wavelength with the data from another wavelength. Extensions of data exchange would expect the data swapping between different time-slots in the time domain (e.g., optical time-division multiplexed (OTDM) network), different polarization states in the polarization domain (e.g., polarization-multiplexed (pol-muxed) network), and different "twisted" light beams carrying different orbital angular momentum (OAM) values in the phase front domain (e.g., OAM-multiplexed network). Moreover, the recently increasing interest of advanced modulation formats [16, 17] would require the data exchange to be available for different modulation formats, such as on-off keying (OOK), differential phase-shift keying (DPSK), differential quadrature phase-shift keying (DQPSK), pol-muxed, etc.

The emergence of nonlinear optics has triggered increased interest and paved a potential way to develop optical signal processing in high-speed optical networks [18, 19]. Optical nonlinearities (e.g. $\chi^{(2)}$ and $\chi^{(3)}$), including difference-frequency generation (DFG) [20, 21], cascaded sum- and difference-frequency generation (cSFG/DFG) [22-26], degenerate/non-degenerate four-wave mixing (FWM) [27-47], and Kerr-induced nonlinear polarization rotation [48-50], are potentially suitable candidates to enable data exchange. In some cases, simple linear optics may also provide an alternative approach to facilitating data exchange [51, 52]. To fulfill the rapid development of high-speed large-capacity optical communications with emerging multiplexing/demultiplexing techniques and advanced modulation formats, as shown in Fig. 1, a laudable goal would be to achieve robust data exchange in different degrees of freedom (wavelength, time, polarization, phase front), for different modulation formats (OOK, DPSK, DQPSK, pol-muxed), and at different granularities (entire data, groups of bits, tributary channels).

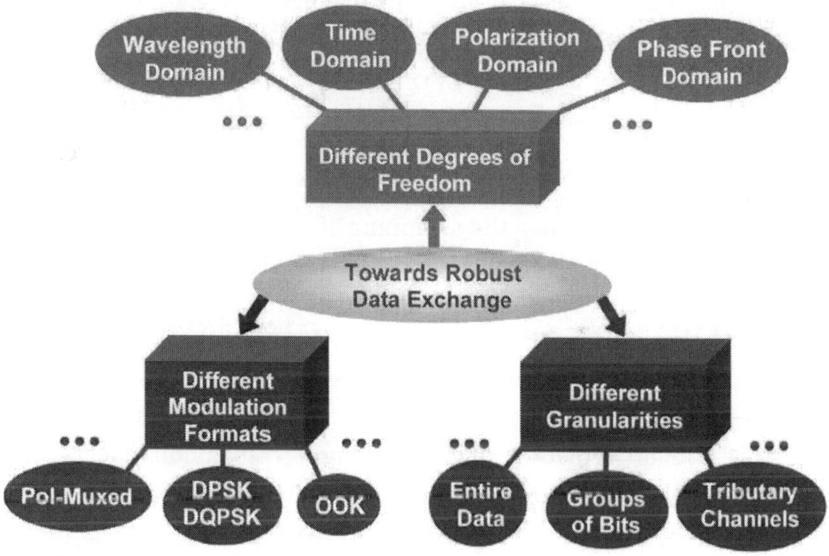

Figure 1. Schematic illustration of robust data exchange.

In this chapter, we tend to provide a comprehensive review of research works towards robust data exchange using various optical nonlinearities [22-50, 53] and simple linear optics [51, 52]. Several kinds of optical nonlinearities are employed: (1) cSFG/DFG in a periodically poled lithium niobate (PPLN) waveguide; (2) non-degenerate FWM in a highly nonlinear fiber (HNLF); (3) bidirectional degenerate FWM in an HNLF; (4) Kerr-induced nonlinear polarization rotation in an HNLF; (5) Conversion-dispersion-based tunable delays. We start with a conceptual description of data exchange followed by state-of-the-art results.

2. CONCEPT OF DATA EXCHANGE

Robust data exchange in the wavelength, time, polarization and phase front domains might be valuable for superior network performance. As an example, a desirable goal of data exchange would be to efficiently utilize nonlinearities in the wavelength domain, such that the data between two different wavelengths can be "exchanged", i.e., swapped, using nonlinear processes in a single device [54]. Figure 2(a) illustrates the basic concept of data exchange in the wavelength domain (wavelength exchange/interchange), which is a wavelength-domain data manipulation enabling the swapping of data between two different wavelengths. One straightforward way, as shown in Fig. 2 (b), is to use two separate wavelength converters (WCs) with one performing the wavelength conversion from signal A (Sig. A) to signal B (Sig. B), and the other from signal B to signal A. Towards single-device operation, one simple way of data exchange in the wavelength domain is to explore the combined signal depletion and wavelength conversion effects in a nonlinear device including a piece of HNLF or a PPLN waveguide [55-58]. Non-degenerate FWM ($\chi^{(3)}$) in an HNLF [29-45] and cascaded second-order nonlinearities ($\chi^{(2)} : \chi^{(2)}$) in a PPLN waveguide [22-26] are potential choices to realize such data exchange. As shown in Fig. 2 (c), due to the signal depletion and wavelength conversion effects, the data carried by signal A is consumed and converted to the wavelength of signal B and vice versa. This enables single-device-based data exchange in the wavelength domain. Similar concepts of data exchange in the time, polarization and phase front domains are also available enabled by various optical nonlinearities or linear optics.

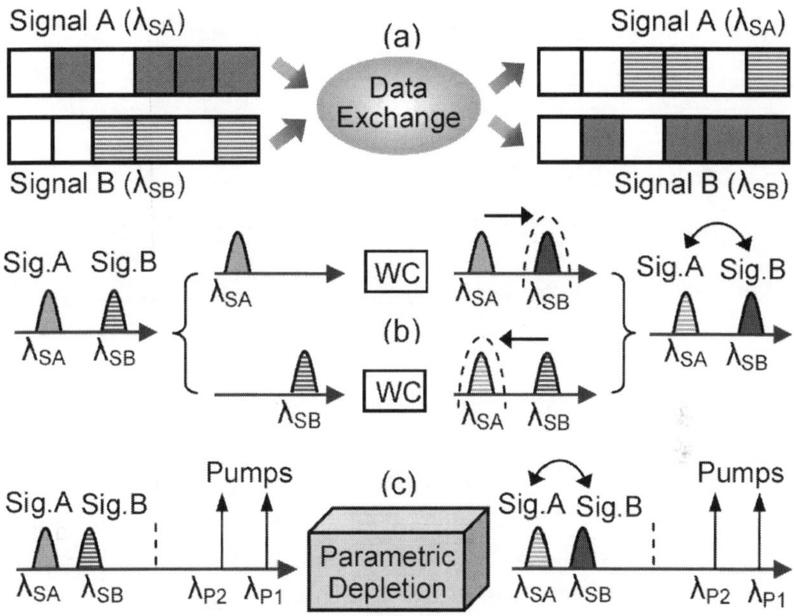

Figure 2. (a) Concept of data exchange in the wavelength domain. (b) Data exchange by two separate wavelength converters (WCs). (c) An example of data exchange by signal depletion and wavelength conversion in a single nonlinear device.

3. RECENT ADVANCES FOR ROBUST DATA EXCHANGE

3.1. Data exchange using cSFG/DFG in a single PPLN waveguide [22-26]

As depicted in Fig. 2(c), data exchange based on signal depletion and wavelength conversion of cSFG/DFG involves two signals and two pumps, which can be described by the coupled-mode equations. To better understand the single-PPLN-based data exchange, under the slowly varying amplitude approximation, we can derive the following analytical solutions to the complex amplitudes of signal A ($A_{SA}(L)$) and signal B ($A_{SB}(L)$) after data exchange [22]

$$A_{SA}(L) = A_{SA}(0) + \omega_{SA}\omega_{SF}\kappa_1 \frac{1}{M^2} A_{P1}^*(0)[\kappa_1 A_{P1}(0)A_{SA}(0) + \kappa_2 A_{P2}(0)A_{SB}(0)][\cos(ML) - 1] \quad (a)$$

$$A_{SB}(L) = A_{SB}(0) + \omega_{SB}\omega_{SF}\kappa_2 \frac{1}{M^2} A_{P2}^*(0)[\kappa_1 A_{P1}(0)A_{SA}(0) + \kappa_2 A_{P2}(0)A_{SB}(0)][\cos(ML) - 1] \quad (b)$$

(1)

where $M = \sqrt{\omega_{SA}\omega_{SF}\kappa_1^2 P_{P1}(0) + \omega_{SB}\omega_{SF}\kappa_2^2 P_{P2}(0)}$. $A_{SA}(0)$, $A_{SB}(0)$, $A_{P1}(0)$ and $A_{P2}(0)$ are the input complex amplitudes of signal A, signal B, pump 1 and pump 2, respectively. $P_{P1}(0)$ and $P_{P2}(0)$

are the input power of pump 1 and pump 2. κ_1 (κ_2) refers to the coupling coefficient of the second-order nonlinear interaction involving signal A (signal B) and pump 1 (pump 2). ω_{SA}, ω_{SB}, and ω_{SF} are the angular frequencies of signal A, signal B and sum-frequency (SF) wave, respectively. L is the waveguide length.

When ignoring the initial pump phase and setting the same power for two input pumps, we can further simplify Eqs. (1a)(1b) as follows

$$A_{SA}(L) = \frac{\cos(ML) + 1}{2} A_{SA}(0) + \frac{\cos(ML) - 1}{2} A_{SB}(0) \quad (a)$$

$$A_{SB}(L) = \frac{\cos(ML) - 1}{2} A_{SA}(0) + \frac{\cos(ML) + 1}{2} A_{SB}(0) \quad (b)$$

(2)

When satisfying the following relationship written by

$$ML = (2N + 1)\pi, \quad N = 0,\ 1,\ 2,\ 3 \cdots\cdots \tag{3}$$

we can obtain

$$\begin{aligned} A_{SA}(L) &= -A_{SB}(0) \quad \text{(a)} \\ A_{SB}(L) &= -A_{SA}(0) \quad \text{(b)} \end{aligned} \tag{4}$$

From Eq. (4) it can be clearly seen that data exchange between signal A and signal B is achieved under the exchange condition governed by Eq. (3). In particular, beyond the data exchange for OOK signal, the complex relationship in Eq. (4) also implies the modulation-format-transparency characteristic of PPLN-based data exchange.

Following the similar principle of PPLN-based data exchange using signal depletion and wavelength conversion of cSFG/DFG, we can further perform robust data exchange functions, including time- and channel-selective data exchange between WDM channels [23, 24] and low-speed tributary channel exchange of high-speed OTDM signals [25, 26].

The conceptual diagram of the proposed single-PPLN-based time- and channel-selective data exchange between WDM channels is illustrated in Fig. 3 [23, 24]. Multiple WDM channels (S1-S4) and two synchronized gated pumps (PA, PB) are coupled into a PPLN waveguide, in which cSFG/DFG processes take place. The wavelength selectivity of the quasi-phase matching (QPM) condition allows selection of channels for data exchange by proper choice of the two pump wavelengths. For proper

QPM of both cSFG/DFG processes, the two pump wavelengths are nearly symmetric to the two exchanged data wavelengths with respect to the QPM wavelength. For instance, as illustrated in Fig. 3, within the gated pump pulse duration, PB mixes with S1 to produce an SF wave through the sum-frequency generation (SFG) process. Meanwhile, the SF wave interacts with PA to generate a new idler at the wavelength of S2 by the subsequent difference-frequency generation (DFG) process. During such nonlinear interactions, S1 can be depleted, and converted to S2 by means of proper control of the pump powers. Similarly, PA and S2 participate in the SFG process to create an SF wave, which simultaneously interacts with PB to yield an idler at the wavelength of S1 via the DFG process. Thus, S2 can also be consumed with its data copied onto S1. Consequently, it is expected to implement optical data exchange between S1 and S2 without the use of additional spectrum and touching other channels. Note that time- and channel-selective data exchange in specific time-slots (groups of bits) and between selective WDM channels can be accomplished by appropriately choosing the gated pump pulse duration and adjusting the pump wavelengths.

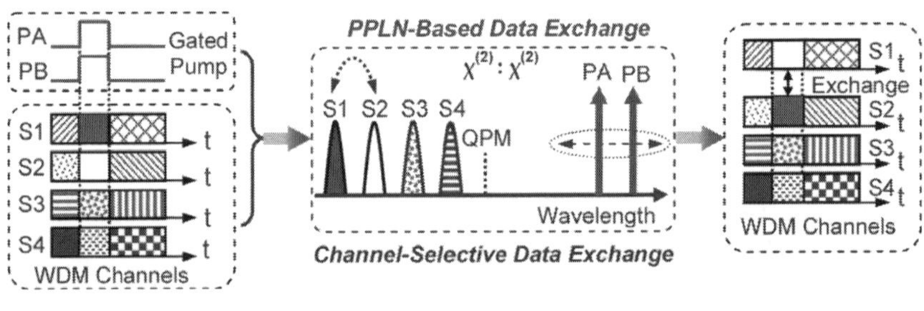

Figure 3. Concept and principle of single-PPLN-based time- and channel-selective data exchange between WDM channels.

We first demonstrate the data exchange between two 10-Gbit/s signals. Two gated pumps with a duty cycle of 1/127 and a pulse duration of ~3.2 ns are employed. The average power of each signal and peak power of each pump coupled into the PPLN waveguide are about 4 mW and 1 W, respectively. Figure 4 displays the observed temporal waveforms and eye diagrams of data exchange. The time-slots between the two straight lines correspond to the gated pump pulse duration, in which data exchange occurs. When S1 and the two pumps are on while S2 is off, the data of S1 within the gated pump pulse duration is depleted (a2) and converted to the wavelength of S2 (b3). Similarly, we can also observe the depletion of S2 (b2) and the conversion from S2 to S1 (a3) by switching S1 off and S2 on. In the case of simultaneously turning on the two signals and the two pumps, it is found that groups of bits data exchange between the two signals (S1 to S2: (b4), S2 to S1: (a4)) within the gated pump pulse duration is successfully achieved.

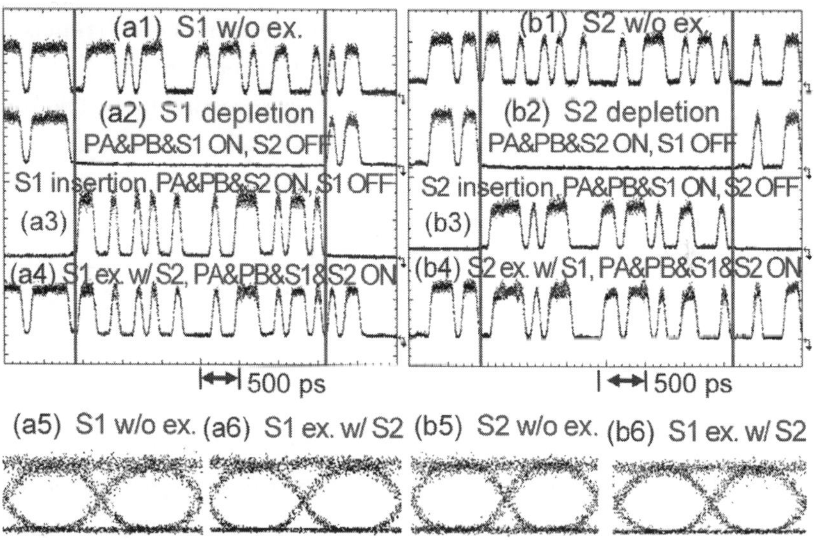

Figure 4. Measured (a1-a4)(b1-b4) temporal waveforms and (a5)(a6)(b5)(b6) eye diagrams of 10-Gbit/s groups of bits data exchange.

We further demonstrate the single-PPLN-based channel-selective data exchange for multiple WDM channels at 40 Gbit/s. Four WDM channels (S1: 1535.5 nm, S2: 1539.4 nm, S3: 1543.3 nm, S4: 1547.2 nm) are employed in the experiment. It is possible to perform a channel-selective data exchange by simply tuning the wavelength of the two pumps. Figure 5 displays the measured typical eye diagrams and bit-error rate (BER) performance for channel-selective data exchange between WDM channels. The power penalty of 40-Gbit/s channel-selective data exchange is estimated to be less than 4 dB at a BER of 10^9.

Figure 6 illustrates the concept and principle of single-PPLN-based tributary channel exchange between two WDM high-speed OTDM signals [25, 26]. A PPLN waveguide is employed as the nonlinear device to perform the tributary channel exchange. Two WDM high-speed signals (S1, S2) each consisting of many low-speed time-division multiplexed tributary channels (e.g., 16 10-Gbit/s tributary channels for 160-Gbit/s signal), together with two synchronized subrate clock (e.g., 10 GHz) pumps, are launched into the PPLN waveguide for the tributary channel exchange. The wavelengths of two signals and two pumps are properly arranged to be symmetric (S1&P1, S2&P2) with respect to the QPM wavelength of PPLN. Inside the PPLN waveguide, two signals and two pumps participate in the cSFG/DFG nonlinear interactions, in which the photons of S1 (S2) and P1 (P2) are annihilated to produce the photons of SF wave, which are simultaneously consumed to generate the photons of S2 (S1) and P2 (P1). Due to the signal depletion and wavelength conversion effects, with the proper adjustment of pump powers, S1 can be depleted and converted to S2. Similarly, S2 can be extinguished to generate S1. As a result, data exchange between two signals (S1, S2) can be implemented. In particular, by exploiting two synchronized subrate (e.g., 10 GHz) clock pumps which are time aligned to one of the tributary channels of two WDM high-speed OTDM signals (e.g., 160 Gbit/s), it is possible to achieve the tributary channel exchange (e.g., 10 Gbit/s) between two WDM high-speed OTDM signals (e.g., 160 Gbit/s). As an example shown in Fig. 6, the tributary channel i (Ch. i) of two WDM high-speed OTDM signals is exchanged by using the signal depletion and wavelength conversion effects of cSFG/DFG in a PPLN waveguide.

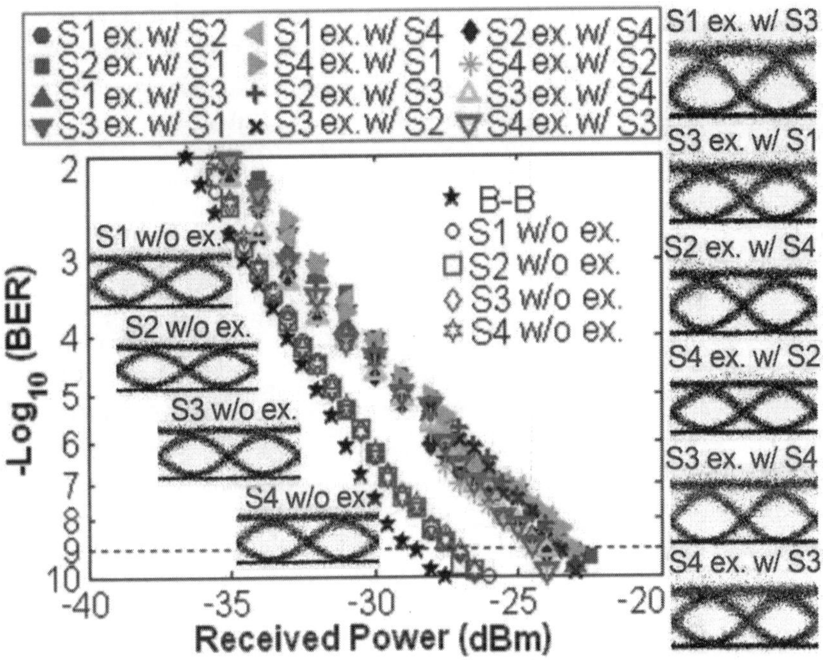

Figure 5. Measured eye diagrams and BER performance of 40-Gbit/s time- (groups of bits) and channel-selective data exchange between four WDM channels.

Figure 7 displays the eye diagrams for tributary channel exchange (Ch.1) measured by an optical sampling scope. Two 10-GHz clock pumps are time aligned to the tributary Ch.1 of two 160-Gbit/s signals. When the two pumps and S1 are present while S2 absent, Ch.1 of S1 is depleted and converted to the Ch.1 of S2 with the proper adjustment of pump powers and polarization states due to the signal depletion and wavelength conversion effects. Similarly, as the two pumps and S2 are turned on while S1 off, Ch.1 of S2 is extinguished with its data information copied onto the Ch.1 of S1. In the presence of two 10-GHz pumps and both two 160-Gbit/s signals, Ch.1 of S2 is exchanged to the Ch.1 of S1. Meanwhile, Ch.1 of S1 is swapped to the Ch.1 of S2, resulting in the implementation of 10-Gbit/s tributary channel exchange between two 160-Gbit/s signals. Moreover, it is convenient to further perform the 10-Gbit/s tributary exchange for all 16 tributary channels of two 160-Gbit/s signals simply by time shifting the 10-GHz clock pumps to be aligned with the corresponding tributary channel of interest.

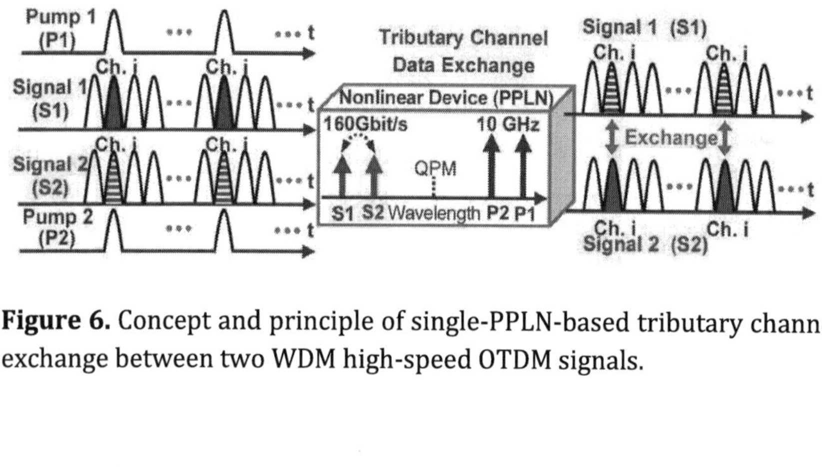

Figure 6. Concept and principle of single-PPLN-based tributary channel exchange between two WDM high-speed OTDM signals.

Figure 8 depicts power penalties at a BER of 10^{-9} of tributary exchange between two 160-Gbit/s signals for all 16 tributary channels. During the tributary channel exchange between two 160-Gbit/s signals, the average power penalty and the fluctuation of 16 tributary channels is around 3.7 and 1.1 dB for S1 (S2 to S1) and 3.9 and 1.1 dB for S2 (S1 to S2).

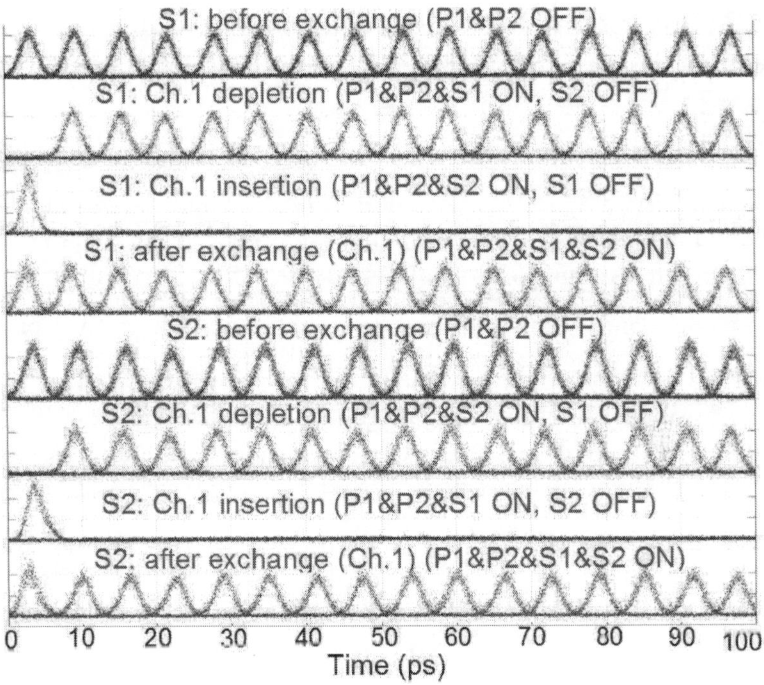

Figure 7. Measured eye diagrams for the tributary channel exchange (Ch. 1).

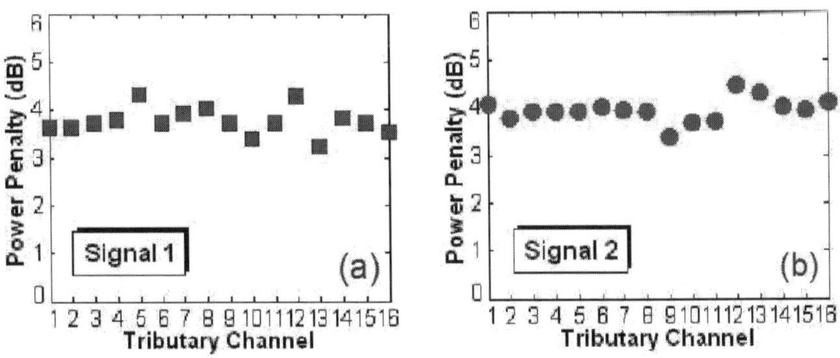

Figure 8. Power penalties of tributary exchange for 16 tributary channels. (a)(c) Signal 1. (b)(d) Signal 2.

3.2. Modulation-format-transparent data exchange using non-degenerate FWM in an HNLF [38-41]

In addition to cSFG/DFG $(\chi^{(2)}:\chi^{(2)})$ in a PPLN waveguide [22-26], signal depletion and wavelength conversion of non-degenerate FWM ($\chi^{(3)}$) in an HNLF can also enable the data exchange [29-41]. As shown in Fig. 9(a), when signal 1 (S1: λ_{S1}) and two continuous-wave (CW) pumps (P1: λ_{P1}, P2: λ_{P2}) are sent through the HNLF with S1 and P1 set symmetrically with respect to the zero-dispersion wavelength (ZDW) of the HNLF, S1 and P1 photons are consumed to produce photons of signal 2 (S2: λ_{S2}) and P2 during the non-degenerate FWM process. Thus the depletion of S1 is expected with its data information transparently copied onto a newly generated S2. Similarly, as shown in Fig. 9(b), the depletion of S2 accompanied by the generation of S1 can be achieved as S2 and two pumps are launched into the HNLF. As shown in Fig. 9(c), in the presence of two signals and two pumps at the input of HNLF with S1(S2) and P1(P2) symmetric relative to the ZDW of the HNLF, S1(S2) can be extinguished and converted to S2(S1), resulting in the implementation of data exchange between S1 and S2.

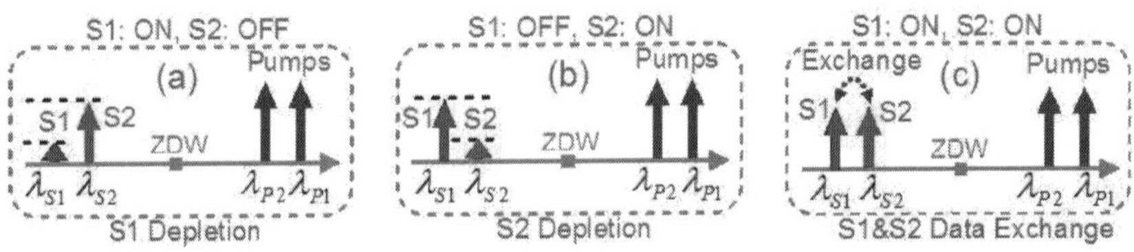

Figure 9. Concept and principle of non-degenerate FWM-based signal depletion and data exchange. (a) S1 depletion. (b) S2 depletion. (c) S1 & S2 data exchange.

For non-degenerate FWM-based data exchange, pump phase modulation is adopted in the experiment to suppress the stimulated Brillouin scattering (SBS) effect so that pump power can be efficiently utilized. Previous works of non-degenerate FWM-based data exchange have been reported for OOK signals [29-37], which are not affected by the phase modulation of two pumps. Shown in Fig. 10 is an example of data exchange (i.e., wavelength exchange) for 10-Gbit/s non-return-to-zero (NRZ) signals [32]. In order to perform phase-transparent data exchange for DPSK/DQPSK signals, it is desired that non-degenerate FWM-based data exchange has the characteristic of modulation-format-transparency.

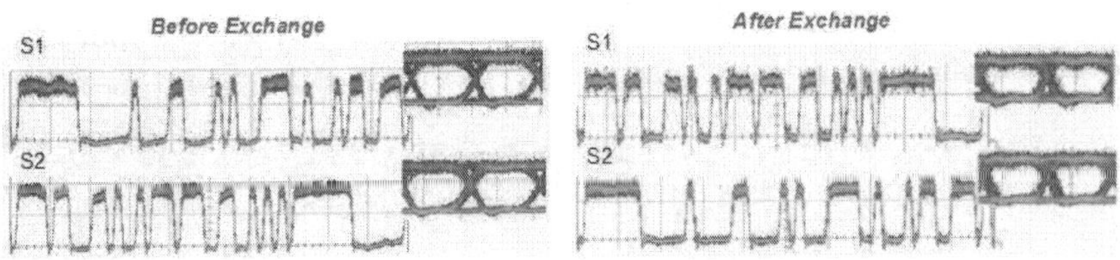

Figure 10. Results of data exchange (wavelength exchange) for 10-Gbit/s NRZ signals [32].

Under the non-depletion approximation, we derive the analytical solutions for the non-degenerate FWM involving two signals and two pumps written as [39]

$$A'_{SA}=\left\{A_{SA0}\left[\cos(gz)-\frac{ik\sin(gz)}{2g}\right]+A_{SB0}\frac{2i\gamma}{g}A^{*}_{P10}A_{P20}\sin(gz)\right\}e^{i\left[2\gamma(P_{10}+P_{20})+\frac{k}{2}\right]z} \quad \text{(a)}$$

$$A'_{SB}=\left\{A_{SA0}\frac{2i\gamma}{g}A_{P10}A^{*}_{P20}\sin(gz)+A_{SB0}\left[\cos(gz)+\frac{ik\sin(gz)}{2g}\right]\right\}e^{i\left[2\gamma(P_{10}+P_{20})-\frac{k}{2}\right]z} \quad \text{(b)}$$

(5)

where $g=\sqrt{4\gamma^2 P_{10}P_{20}+k^2/4}$ and $k=\Delta\beta+\gamma(P_{10}-P_{20})$ and =$\Delta\beta$+γ(P 10 −P 20) k=Δβ+γ(P10−P20)

are constants related to the pump powers (P_{10}, P_{20}), nonlinear coefficient (γ) and phase mismatching $(\Delta\beta)$. A_{SA0}, A_{SB0}, A_{P10} and A_{P20} are the complex amplitudes of input signals (SA, SB) and pumps (P1, P2) containing both amplitude and phase information. $A_{SA}^{'}$ and $A_{SB}^{'}$ are the complex amplitudes of output signals (SA, SB) after the data exchange. Under the exchange condition of phase =0) and $gz=(N+1/2)\pi$ $(N=0,\ 1,\ 2,\ ...)$

enabled by the proper adjustment of pump powers, we can further simplify Eqs. (5a)(5b) as follows

$$A_{SA}^{'}=\pm A_{SB0}\frac{2i\gamma}{g}A_{P10}^{*}A_{P20}e^{i2\gamma(P_{10}+P_{20})z} \quad \text{(a)}$$
$$A_{SB}^{'}=\pm A_{SA0}\frac{2i\gamma}{g}A_{P10}A_{P20}^{*}e^{i2\gamma(P_{10}+P_{20})z} \quad \text{(b)}$$

(6)

Note that Eqs. (6a)(6b) indicate the linear relationship of complex amplitude between the output and input signals $(A_{SA}^{'}\propto A_{SB0},\ A_{SB}^{'}\propto A_{SA0})$, implying the implementation of phase-transparent optical data exchange. We can further obtain the corresponding phase relationships of $\phi_{SA}^{'}=\phi_{SB}+\phi_{P2}-\phi_{P1}$ and $\phi_{SB}^{'}=\phi_{SA}+\phi_{P1}-\phi_{P2}$. Remarkably, the pump phase $(\phi_{P1}-\phi_{P2}\neq 0)$ to the exchanged signals does not impact on the OOK data exchange but could cause severe degradation on the DPSK/DQPSK data exchange. Fortunately, according to the deduced phase relationships, it is possible to cancel the pump phase transfer by applying the precisely identical phase modulation to the two (i.e., $\phi_{P1}=\phi_{P2}$), which makes it applicable to implement the data exchange of DPSK/DQPSK signals [38-41].

A 1-km piece of HNLF is adopted in the experiment, which has a nonlinear coefficient of 9.1 $W^{-1}\cdot km^{-1}$, a ZDW of ~1552 nm, and a fiber loss of 0.45 dB/km. To suppress SBS, identical phase modulation is applied to the two pumps using a single phase modulator (PM) driven by a 10-Gbit/s pseudo-random binary sequence (PRBS). According to Eqs. (6a)(6b), the precisely identical phase modulation of the two pumps could be canceled in the output signals after data exchange.

We demonstrate the phase-transparent data exchange between two 100-Gbit/s 2^7-1 PRBS return-to-zero DQPSK (RZ-DQPSK) signals (S1: signal 1, S2: signal 2) [40, 41]. Figure 11(a) displays the measured temporal waveforms of the demodulated in-phase (Ch. I) and quadrature (Ch. Q) components for the 100-Gbit/s RZ-DQPSK data exchange. It can be clearly observed that the data information carried by two 100-Gbit/s RZ-DQPSK signals is successfully swapped after the non-degenerate FWM-based data exchange. Figure 11(b) and (c) plot BER curves for the 100-Gbit/s RZ-DQPSK data exchange. Less than 1.2-dB power penalty at a BER of 10^{-9} is obtained for the 100-Gbit/s RZ-DQPSK wavelength conversion (WC) with only one signal (S1 or S2) present. Less than 5-dB power penalty at a BER of 10^{-9} is observed for the 100-Gbit/s RZ-DQPSK data exchange. The extra power penalty of data exchange compared to wavelength conversion could be ascribed to the beating effect between the newly converted signal and the original residual signal.

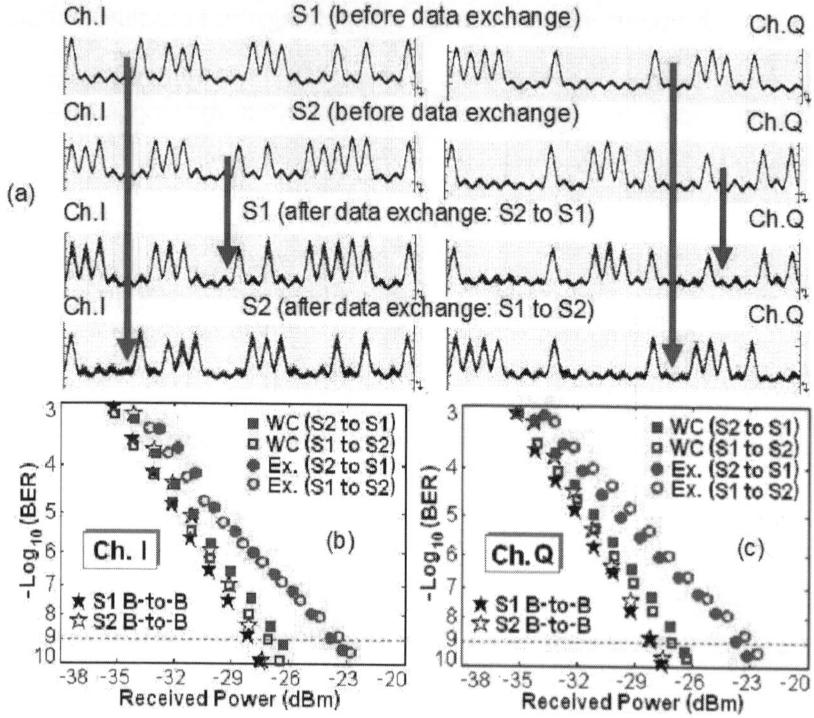

Figure 11. Measured (a) demodulated waveforms and (b)(c) BER curves for 100-Gbit/s RZ-DQPSK data exchange.

3.3. Multi-channel data exchange using bidirectional degenerate FWM in an HNLF [42-45]

The aforementioned signal depletion and wavelength conversion based schemes with two pumps enable the two-channel data exchange [22-26, 29-41]. However, the extended applications to simultaneous multi-channel data exchange might be limited. A laudable goal would be to explore the data exchange between multi-channel signals.

Figure 12 illustrates the concept and principle of multi-channel data exchange [42, 43]. Degenerate FWM with a single CW pump is utilized. Four-channel DQPSK signals (S1-S4) are symmetric with respect to the

CW pump. Simultaneous data exchange between S1 and S4 as well as S2 and S3 is expected. In general, such exchange function is not applicable with the unidirectional degenerate FWM in a single HNLF since the newly converted signals cannot be separated from the original signals. A potential solution is to explore the bidirectional degenerate FWM in a single HNLF assisted by optical filtering. As shown in Fig. 12, for the input four-channel signals (S1-S4), the filtered S1, S2 and CW pump are sent to HNLF from the left side, yielding S4 and S3 via degenerate FWM. The newly generated S4 and S3 are selected at the right side of HNLF while the original S1, S2 and CW pump are blocked. Meanwhile, the filtered S3, S4 and CW pump are fed into HNLF from the right side, producing S2 and S1 by degenerate FWM. The newly converted S2 and S1 are selected at the left side of HNLF while the original S3, S4 and CW pump are removed. As a consequence, simultaneous four-channel data exchange (S1&S4, S2&S3) can be achieved using bidirectional FWM in a single HNLF assisted by optical filtering. The combined S1-S4 from both sides of HNLF are the output four-channel signals after data exchange. Note that the in-phase (Ch. I) and quadrature (Ch. Q) components of DQPSK signals are swapped after data exchange due to the phase-conjugation characteristic of degenerate FWM.

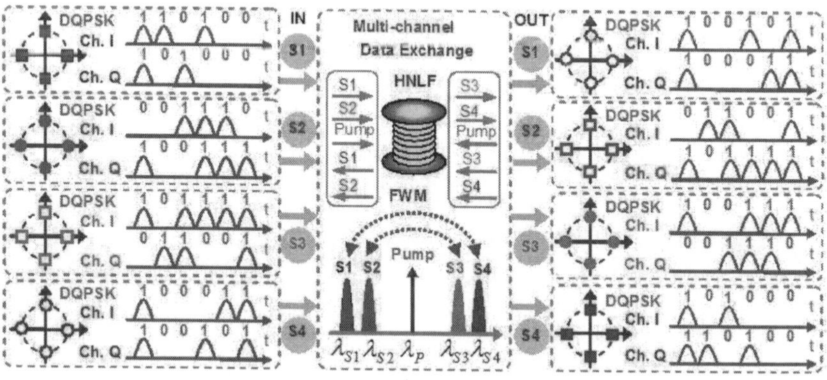

Figure 12. Concept and principle of simultaneous multi-channel DQPSK data exchange.

The proposed simultaneous multi-channel data exchange can be incorporated in a reconfigurable network switching element to enhance the efficiency and flexibility of optical networks. We construct a reconfigurable Tbit/s network switching element using double-pass liquid crystal on silicon (LCoS) technology accompanied by bidirectional degenerate FWM in a single HNLF. We demonstrate the LCoS+HNLF-based 2.3-Tbit/s multi-functional grooming switch which performs simultaneous selective add/drop, switchable data exchange, and power equalization, for 23-channel 100-Gbit/s RZ-DQPSK signals [44, 45].

ITU-grid-compatible 23-channel (from S1: 1531.12 nm to S23: 1566.31 nm) 100-Gbit/s RZ-DQPSK signals are employed in the experiment. Figure 13 shows the measured spectrum of the input unequalized 23-channel 100-Gbit/s RZ-DQPSK signals with a power fluctuation of ~9.1 dB. Shown in the insets are typical balanced eyes for the in-phase (Ch. I) and quadrature (Ch. Q) components.

Shown in Fig. 14 is the measured spectrum and balanced eyes after grooming switch with power equalization (<1 dB) for all 23 channels (input unequalization: ~9.1 dB), two-channel add/drop for S6 and S7, and simultaneous six-channel data exchange (S10, S11, S12, S21, S22, S23). The inset of Fig. 14 depicts the spectrum of dropped S6 and S7. The BER performance is plotted in Fig. 15 and power penalties less than 5 dB for six-channel data exchange are observed at a BER of 10^{-9}.

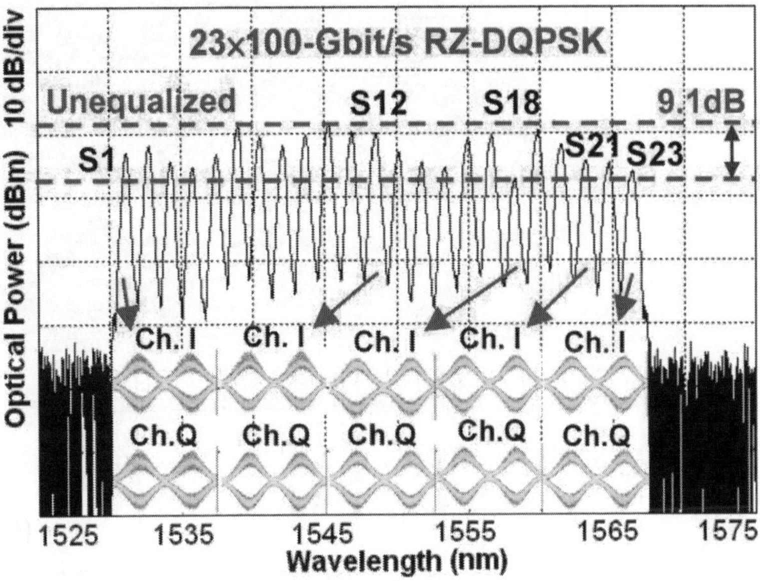

Figure 13. Measured spectrum and balanced eyes for input unequalized 23-channel 100-Gbit/s RZ-DQPSK signals.

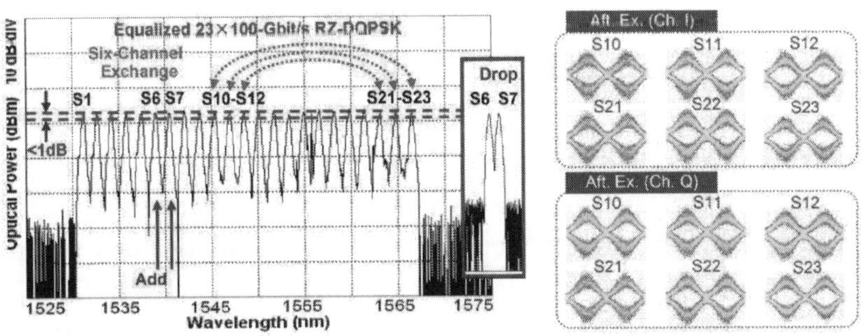

Figure 14. Measured spectrum and balanced eyes after multi-functional grooming switch (S6, S7: add/drop; S10, S11, S12, S21, S22, S23: data exchange; S1-S23: power equalization).

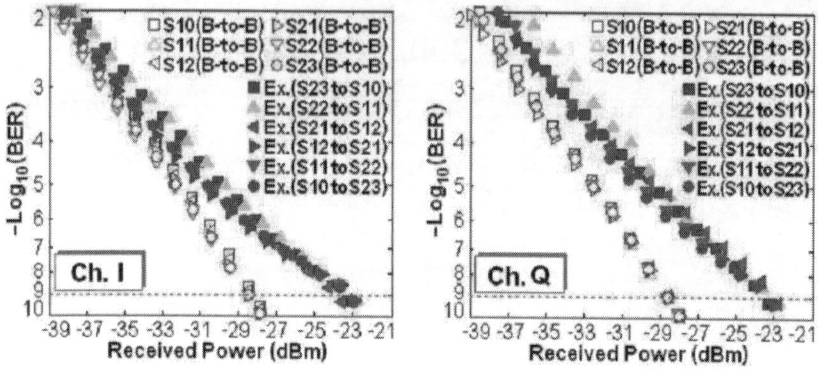

Figure 15. BER curves for simultaneous six-channel data exchange (S10, S11, S12, S21, S22, S23).

3.4. Data exchange between two orthogonal polarizations using kerr-induced nonlinear polarization rotation in an HNLF [48, 49]

In addition to the data exchange in the wavelength and time domains [22-45], it is also possible to perform data exchange between two orthogonal polarizations in the time and polarization domains [48-50]. We experimentally demonstrate the orthogonal tributary channel exchange between two pol-muxed DPSK OTDM data streams by using the Kerr effect-induced nonlinear birefringence in an HNLF [48, 49].

Figure 16 illustrates the concept and principle of the Kerr effect-based orthogonal tributary channel exchange of a pol-muxed DPSK OTDM signal. The strong subrate clock pump is 45° linearly polarized with respect to the two orthogonal polarizations of a pol-muxed DPSK OTDM signal. With the help of proper pump power control, the pump-induced nonlinear birefringence by Kerr effect could bring the selected tributary channel (aligned with the subrate clock pump) to a 90° polarization rotation for both of the two orthogonal polarizations of the pol-muxed signal, leading to the orthogonal tributary channel exchange when the pump is present. Other unselected orthogonal tributary channels with the pump absent will not experience the nonlinear polarization rotation and hence will be untouched. In addition, simply by shifting the subrate clock pump to be aligned with the tributary channel of interest, it is possible to implement orthogonal tributary channel exchange for all tributary channels of the pol-muxed DPSK OTDM signal.

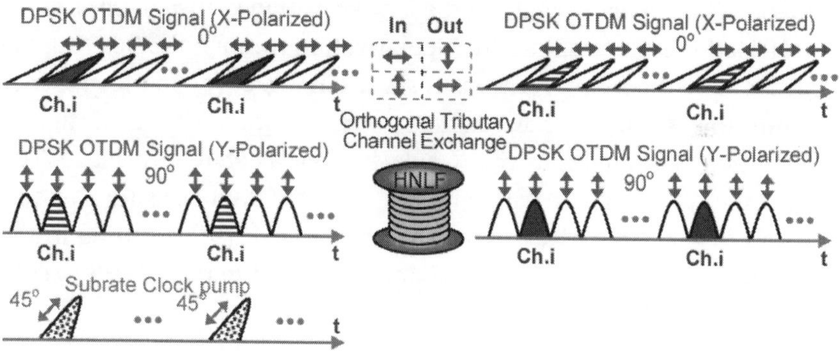

Figure 16. Concept and principle of Kerr effect-based orthogonal tributary channel exchange of a pol-muxed DPSK OTDM signal.

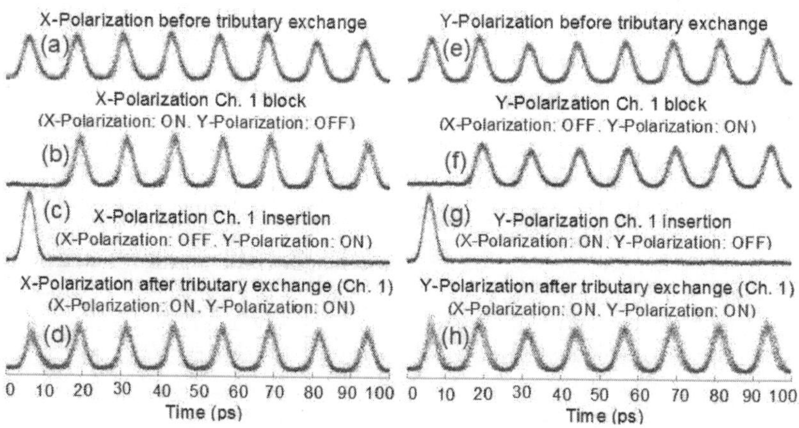

Figure 17. Eye diagrams of orthogonal tributary channel (Ch. 1) exchange of a 160-Gbit/s pol-muxed DPSK OTDM signal.

Figure 17 displays the eye diagrams measured by an optical sampling scope for the typical orthogonal tributary channel (Ch. 1) exchange of a 160-Gbit/s pol-muxed DPSK OTDM signal. As the 10-GHz clock pump is time aligned to tributary channel 1 (Ch. 1) of the X- and Y-polarized DPSK OTDM signal, in the absence of the Y-polarization, as shown in Fig. 17(b), Ch. 1 of the X-polarization is blocked by an X-polarizer after the HNLF due to the 90° rotation from the X- to the Y-polarization. When the Y-polarization is present but the X-polarization is absent, Ch. 1 of the Y-polarization is inserted to the X-polarization through the 90° rotation from the Y- to the X-polarization, as shown in Fig. 17(c). In the presence of both the X- and Y-polarizations, the tributary Ch. 1 of the Y-polarization is changed to the X-polarization, as shown in Fig. 17(d). Meanwhile, the original tributary Ch. 1 of the X-polarization is also changed to the Y-polarization, as shown in Fig. 17(h), resulting in the orthogonal tributary channel exchange of a pol-muxed DPSK OTDM signal.

Figure 18 plots the power penalties of the orthogonal tributary exchange for 8 tributary channels. Less than 4-dB power penalty at a BER of 10^{-9} is obtained for all 8 tributary channels with a fluctuation of <1.5 dB.

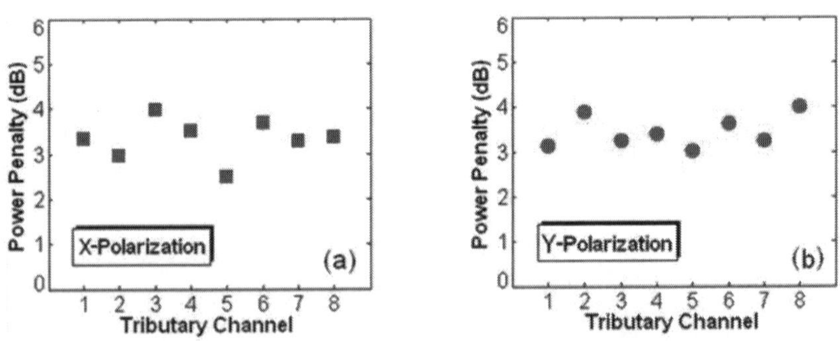

Figure 18. Power penalties of orthogonal tributary exchange for 8 tributary channels.

3.5. Time-slot-exchange using conversion-dispersion-based tunable delays [46, 47]

The demonstrated data exchange of groups of bits or tributary channels manipulates data in multiple degrees of freedom, such as time- (groups of bit) and channel-selective data exchange between WDM channels [23, 24], tributary channel exchange between two WDM high-speed OTDM signals [25, 26], and orthogonal tributary channel exchange of a pol-muxed OTDM signal [48-50]. Another important traffic grooming function, known as time-slot exchange or time-slot interchange, is to manipulate data only in the time domain to enable contention resolution and increase throughput efficiency in time-based networks. Time-slot exchange, occurring on the bit or packet level, can afford the network enhanced flexibility. For packet-switched networks, exchanging full data packets in the time domain requires optical delays that are tunable.

Figure 19 shows concept and principle of conversion-dispersion-based time-slot exchange of two separate packets in the time domain [46]. Three clocked pumps $(\lambda_{P1}, \lambda_{P2}, \lambda_{P3})$ are fed into an HNLF, along with a packetized input signal (λ_S) located near the ZDW of HNLF. Degenerate FWM between the clocked pumps and signal generates replicas of the input signal at new converted $(\lambda_1, \lambda_2, \lambda_3)$ which contain only the information of the input signal at times when the clocked pumps are on. The three pumps are clocked to convert: (i) only packet A to λ_1, (ii) all information but packets A and B to λ_2, and (iii) only packet B to λ_3. The three converted signals $(\lambda_1, \lambda_2, \lambda_3)$ then pass through a dispersion module, such as dispersion compensation fiber (DCF), and experience a wavelength-dependent delay via inter-channel chromatic dispersion. Due to the conversion-dispersion-based tunable delays, packet A is advanced, while packet B is retarded, relative to the reference (all information but packets A and B), resulting in the swapping of packets A and B in the time domain. After the delays, all three converted signals are converted back to the original signal wavelength using a PPLN waveguide followed by the compensation for intra-channel chromatic dispersion.

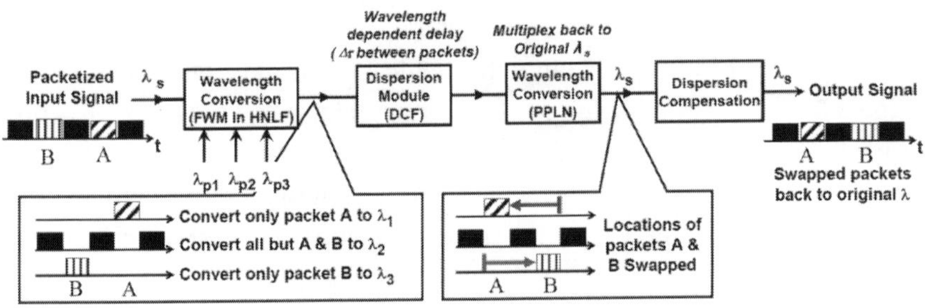

Figure 19. Concept and principle of time-slot exchange of two separate packets using conversion-dispersion-based tunable delays.

Shown in Figs. 20 and 21 are experimental results of time-slot exchange of 40-Gbit/s optical data packets using conversion-dispersion-based tunable delays. Separate 182-bit packets are converted to separate wavelengths, delayed relative to one another using conversion-dispersion-based tunable delays, and then recombined together to achieve new packets with time slots exchanged.

Similarly, time-slot exchange between odd and even data packets is also achievable using conversion-dispersion-based tunable delays [47]. As illustrated in Fig. 22, odd and even data packets are extracted from an input signal via wavelength multicasting with clocked pumps, delayed relative to one another in a dispersion module, and then multiplexed back together using wavelength conversion in a PPLN waveguide followed by dispersion compensation. Shown in Fig. 23 are experimental results of time-slot exchange of 40-Gbit/s odd and even packets. The conversion-dispersion-based tunable delays enable time-slot exchange of variable length optical packets (182 and 288 bits/packet).

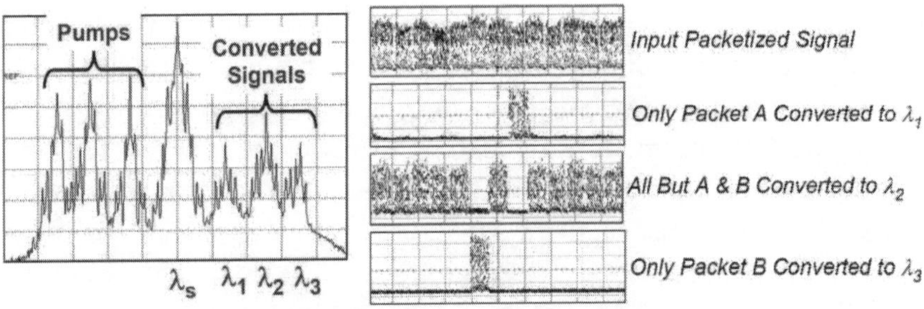

Figure 20. Optical spectra (left) and temporal waveforms (right) of converted signals after multicasting using clocked pumps in HNLF.

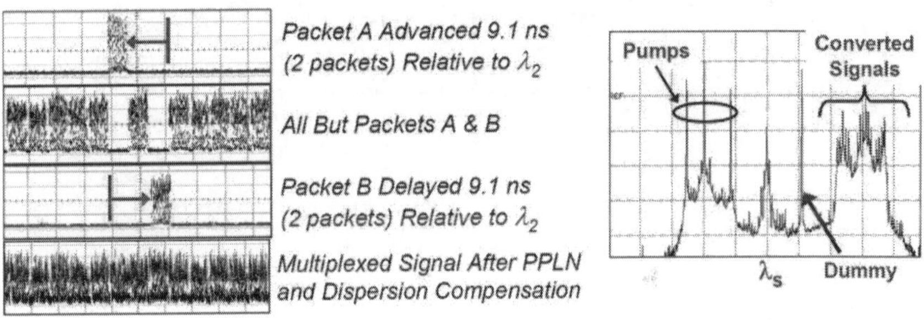

Figure 21. Temporal waveforms of optical signals following delay via dispersion, conversion back to the original signal wavelength and dispersion compensation (left) and optical spectra after PPLN (right).

3.6. Data exchange between “twisted” light beams carrying Orbital Angular Momentum (OAM) [51, 52]

In optical communications, beyond well-known existing degrees of freedom such as wavelength, time and polarization, other degrees of freedom are encouraged to be explored to break “capacity crunch”. For example, OAM which is related to the helical phase front of “twisted” light beams [59-61], can be considered as an additional degree of freedom [62, 63], where the multiplexing of data-carrying OAM beams provides yet another dimension in the ever-continuing effort to increase the capacity and spectral efficiency of communication links [63]. When employing OAM beams to carry data information, a desirable function for flexible data processing would be the data exchange between "twisted" OAM beams.

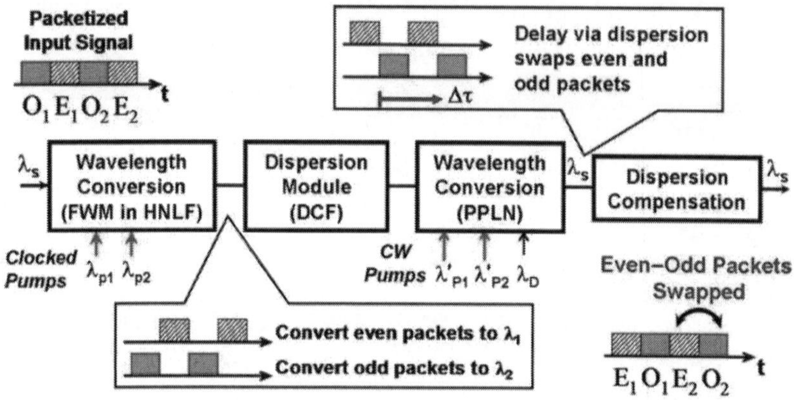

Figure 22. Concept and principle of time-slot exchange of odd and even packets.

Figure 24 shows concept and principle of data exchange between OAM beams [51, 52]. Superposed two OAM beams ($\mathrm{OAM}_{\ell 1\ \ell 1}$, $\mathrm{OAM}_{\ell 2\ \ell 2}$), each carrying different data information (Signal A, Signal B), shine at a reflective-type spatial light modulator (SLM) loaded with a spiral phase mask with a charge of $\ell_R = -(\ell_1 + \ell_2)$. After reflecting off the SLM, this phase mask adds an azimuthal phase term $\exp(i\ell_R\theta)$ to the two OAM beams and converts them into $\mathrm{OAM}_{-\ell 2\ -\ell 2}$ and $\mathrm{OAM}_{-\ell 1\ -\ell 1}$, which are further transformed into $\mathrm{OAM}_{\ell 2\ \ell 2}$ and $\mathrm{OAM}_{\ell 1\ \ell 1}$ due to reflection of the SLM which flips the charge sign. As a result, data exchange between two OAM beams is implemented. For the input of two OAM beams with varied charges, reconfigurable data exchange is available by updating the phase mask loaded into the reflective-type SLM. Shown in Fig. 24 is an example of data exchange between DQPSK-carrying "twisted" OAM beams (OAM_{+8}, OAM_{+6}).

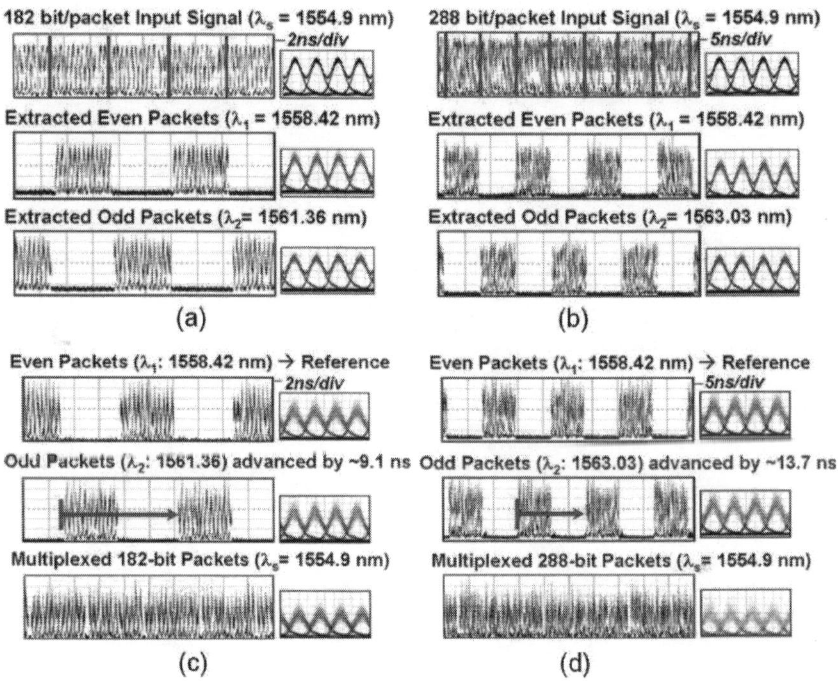

Figure 23. Temporal waveforms of time-slot exchange of 40-Gbit/s odd and even packets with two variable packet lengths. (a)(c) 182 bits/packet. (b)(d) 288 bits/packet.

The measured interferograms (i.e., interference between OAM beams and a reference Gaussian beam), as shown in Fig. 25(a) and (b), indicate that the charges of two OAM beams before exchange are +8 and +6. After exchange, the measured interferograms, as shown in Fig. 25(c) and (d), verify that the charges of two OAM beams after exchange become +6 and +8 (see Ref. 52 for details). Figure 25(e)-(h) show measured interferograms of reconfigurable data exchange between another two OAM beams (OAM_{+10} and OAM_{+6}) by updating the spiral phase mask loaded into the SLM.

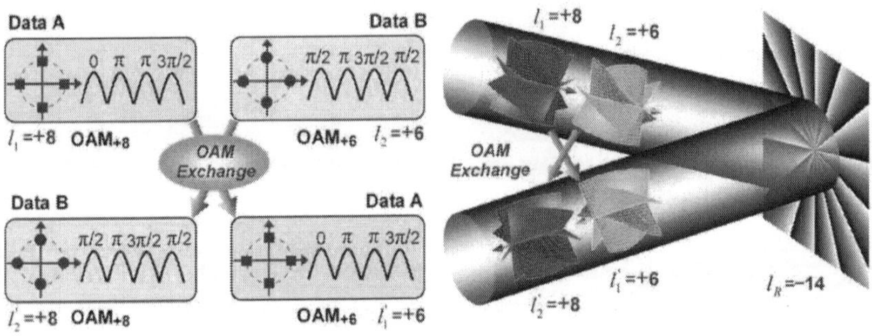

Figure 24. Concept and principle of data exchange between "twisted" OAM beams.

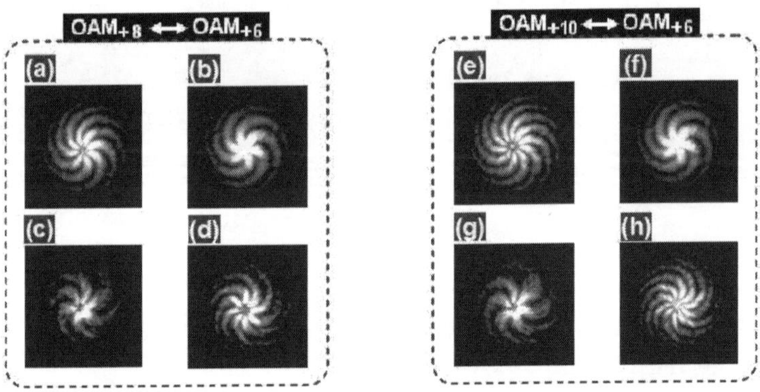

Figure 25. Measured interferograms. (a) OAM_{+8} and (b) OAM_{+6} beams become (c) OAM_{+6} and (d) OAM_{+8} ones after exchange. (e) OAM_{+10} and (f) OAM_{+6} beams become (g) OAM_{+6} and (h) OAM_{+10} ones after exchange.

We measure temporal waveforms and balanced eyes of demodulated in-phase (Ch. I) and quadrature (Ch. Q) components of 100-Gbit/s RZ-

DQPSK signals. As shown in Fig. 26, the observed temporal waveforms confirm the successful implementation of data exchange between two OAM beams (OAM_{+10} and OAM_{+6}). Shown in Fig. 27 are measured BER curves for 100-Gbit/s RZ-DQPSK data exchange between OAM_{+10} and OAM_{+6} beams with power penalty <1.9-dB at a BER of $1e^{-9}$.

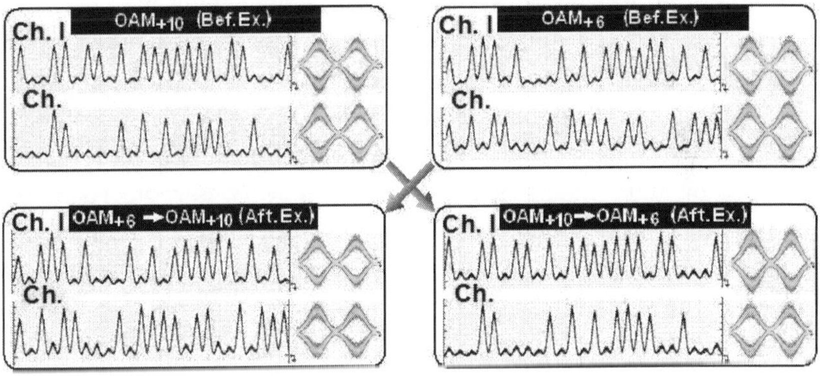

Figure 26. Measured waveforms and balanced eyes of demodulated in-phase (Ch. I) and quadrature (Ch. Q) components of 100-Gbit/s RZ-DQPSK signals for data exchange between OAM_{+10} and OAM_{+6} beams. Bef. Ex., before exchange; Aft. Ex., after exchange.

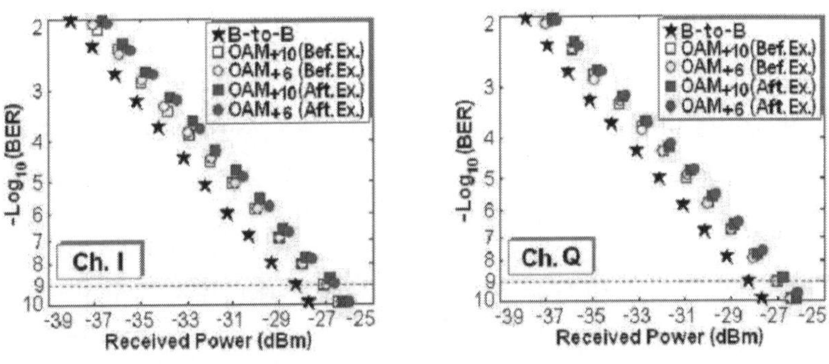

Figure 27. Measured BER curves for 100-Gbit/s RZ-DQPSK data exchange between OAM_{+10} and OAM_{+6} beams.

4. DISCUSSIONS

The demonstrated miscellaneous data exchange functionalities provide great potential for facilitating flexible networks. With future improvement, several aspects could be considered as follows.

1 In practical applications, some supplementary functionalities might be required to construct complete and independent data exchange modules. Taking tributary channel exchange as an example (Figs. 6-8), subrate clock pumps should be synchronized with high-speed OTDM signals. Note that the incoming signals and locally generated pumps are usually independent with each other. Hence, a supplementary functionality of clock recovery is required in real situations to get synchronized subrate clock pumps from incoming signals. Fortunately, various optical clock recovery methods have been developed [64]. In particular, recent promising demonstrations have shown the successful synchronization and sub-clock recovery for ultra-high speed OTDM signals up to 640 Gbit/s [65, 66]. As a consequence, it is possible develop a complete and independent data exchange module by incorporating a synchronization and clock recovery unit.

2 Beyond reported functionalities, data exchange can be further extended in terms of degrees of freedom, modulation formats, and granularities. For example, some additional degrees of freedom have recently attracted increasing interest in high-speed optical fiber communications to break the “capacity crunch”, such as space [67, 68] and mode [69, 70]. A valuable goal would be achieve data exchange in these degrees of freedom. Also, some high-level modulation formats have been used in fiber transmission, such as 16-ary quadrature amplitude modulation (16-QAM), 32-QAM, etc [67, 68, 71]. According to the demonstrated characteristic of modulation-format-transparency, most of the presented data exchange should be, in principle, available for these advanced modulation formats. However, high-level modulation formats show reduced tolerable performance degradation, and therefore accurate

manipulation of amplitude and phase would be expected. Additionally, data exchange with fine granularity in the time domain requires accurate control of time delay, which could be achievable assisted by the fine tuning of conversion-dispersion-based optical delays [7].

3 In addition to PPLN and HNLFs, there would be some other alternative candidates applicable for data exchange, including the use of third-order nonlinearities in semiconductor optical amplifiers (SOAs) [3], As_2S_3 waveguides [19], and silicon waveguides [18].

5. CONCLUSION

In this chapter, we have reviewed recent research efforts towards robust data exchange. Various kinds of optical nonlinearities, i.e., cSFG/DFG in a PPLN waveguide, non-degenerate FWM in an HNLF, bidirectional degenerate FWM in an HNLF, Kerr-induced nonlinear polarization rotation in an HNLF, and conversion-dispersion-based tunable delays, together with simple linear optics, are exploited to enable robust data exchange in different degrees of freedom (wavelength, time, polarization, phase front), for different modulation formats (OOK, DPSK, DQPSK, pol-muxed), and at different granularities (entire data, groups of bits, tributary channels).

First, analytical solutions to the single-PPLN-based data exchange are derived showing the exchange condition. 40-Gbit/s time- (groups of bits) and channel-selective data exchange between four WDM channels is implemented. 10-Gbit/s tributary channel exchange between two WDM 160-Gbit/s OTDM signals is demonstrated. Second, analytical solutions to the non-degenerate FWM-based data exchange are derived indicating the exchange condition and implying the characteristic of modulation-format-transparency. Phase-transparent data exchange (entire data) of 100-Gbit/s RZ-DQPSK signals is demonstrated. Third, a simple approach is proposed to perform simultaneous multi-channel data exchange using bidirectional degenerate FWM in an HNLF. A reconfigurable Tbit/s network switching element is constructed using double-pass LCoS technology, together with bidirectional degenerate FWM in a single HNLF. LCoS+HNLF-based 2.3-Tbit/s (23X100-Gbit/s RZ-DQPSK) multi-functional grooming switch (e.g., simultaneous add/drop, six-channel data exchange, and power equalization) is implemented. Fourth, 10-Gbit/s tributary channel exchange between two orthogonal polarizations of a 160-Gbit/s pol-muxed DPSK OTDM signal is demonstrated based on the Kerr-induced nonlinear polarization rotation. Fifth, time-slot exchange of 40-Gbit/s optical data packets is demonstrated using conversion-dispersion-based tunable delays. Finally, reconfigurable 100-Gbit/s RZ-DQPSK data exchange between "twisted" OAM beams is demonstrated using simple linear optics.

The obtained theoretical and experimental results of data exchange in the wavelength, time, polarization and phase front domains, show that robust data exchange for different modulation formats and at different granularities could potentially enhance the efficiency and flexibility of optical networks.

ACKNOWLEDGEMENTS

We acknowledge Jeng-yuan Yang, Xiaoxia Wu, Scott R. Nuccio, Omer F. Yilmaz, Zahra Bakhtiari, Hao Huang, Xue Wang, Nisar Ahmed, Irfan Fazal, Yan Yan, Yang Yue, Lin Zhang, Yinying Xiao-Li, Bishara Shamee, Yongxiong Ren, Amanda Bozovich, Robert W. Hellwarth, Moshe Tur, Kevin Birnbaum, John Choi, Baris Erkmen and Samuel Dolinar for the helpful discussions, and the generous support of the National Natural Science Foundation of China (NSFC) under grants 61077051, 11274131, 61222502, the Program for New Century Excellent Talents in University (NCET-11-0182), the Defense Advanced Research Projects Agency (DARPA) under contract FA8650-08-1-7820, and the DARPA under InPho (Information in a Photon) program.

REFERENCES

1. Saruwatari M. All-optical signal processing for terabit/second optical transmission. IEEE J. Sel. Top. Quantum Electron. 2000; 6(6): 1363-1374.

2. Yoo SJB Wavelength conversion technologies for WDM network applications. J. Lightwave Technol. 1996; 14(6): 955-966.

3. Chan K, Chan CK, Chen LK, Tong F. Demonstration of 20-Gb/s all-optical XOR gate by four-wave mixing in semiconductor optical amplifier with RZ-DPSK modulated inputs. IEEE Photon. Technol. Lett. 2004; 16(3): 897-899.

4. Wang J, Sun JQ, Zhang XL, Huang DX, Fejer MM. All-optical format conversions using periodically poled lithium niobate waveguides. IEEE J. Quantum Electron. 2009; 45(2): 195-205.

5. Okawachi Y, Sharping JE, Xu C, Gaeta AL. Large tunable optical delays via self-phase modulation and dispersion. Opt. Express 2006; 14(25): 12022-12027.

6. Wang Y, Yu CY, Yan LS, Willner AE, Roussev R, Langrock C, Fejer MM, Sharping JE, Gaeta AL. 44-ns continuously tunable dispersionless optical delay element using a PPLN waveguide with two-pump configuration, DCF, and a dispersion compensator. IEEE Photon. Technol. Lett. 2007; 19(11): 861-863.

7. Nuccio SR, Yilmaz OF, Wu X, Willner AE. Fine tuning of conversion/dispersion based optical delays with a 1 pm tunable laser using cascaded acousto-optic mixing. Opt. Lett. 2010; 35(4): 523-525.

8. Dai Y, Okawachi Y, Turner-Foster AC, Lipson M, Gaeta AL, Xu C. Ultralong continuously tunable parametric delays via a cascading discrete stage. Opt. Express 2010; 18(1): 333-339.

9. Salem R, Foster MA, Turner AC, Geraghty DF, Lipson M, Gaeta AL. Signal regeneration using low-power four-wave mixing on silicon chip. Nature Photonics 2008; 2(1): 35-38.

10. Kataoka N, Sone K, Wada N, Aoki Y, Kinoshita S, Miyata H, Miyazaki T, Onaka H, Kitayama K. Field trial of 640-Gbit/s-throughput, granularity-flexible optical network using packet-selective ROADM prototype. J. Lightwave Technol. 2009; 27(7): 825-832.

11. Wang J, Fu HY, Geng DY, Willner AE. All-optical wavelength-/time-selective switching/dropping/swapping for 100-GHz-spaced WDM signals using a periodically poled lithium niobate waveguide. ECOC2012, paper Th.1.A.5, 2012.

12. Wu XX, Bogoni A, Yilmaz OF, Nuccio SR, Wang J, Willner AE. Eightfold 40-320 Gbit/s phase-coherent multiplexing and 320-40 Gbit/s demultiplexing using highly nonlinear fibers. Opt. Lett. 2010; 35(11): 1896-1898.

13. Brès CS, Boggio JMC, Alic N, Radic S. 1-to-40 10-Gb/s channel multicasting and amplification in wideband parametric amplifier. IEEE Photon. Technol. Lett. 2008; 20(16): 1417-1419.

14. Biberman A, Lee BG, Turner-Foster AC, Foster MA, Lipson M, Gaeta AL, Bergman K. Wavelength multicasting in silicon photonic nanowires. Opt. Express 2010; 18(17): 18047-18055.

15. Hamza HS, Deogun JS. Wavelength-exchanging cross connects (WEX)—a new class of photonic cross-connect architectures. J. Lightwave Technol. 2006; 24(3): 1101-1111.

16. Winzer PJ, Essiambre RJ. Advanced optical modulation formats. Proc. IEEE 2006; 94(5): 952-985.

17. Winzer PJ, Essiambre RJ. Advanced modulation formats for high-capacity optical transport networks. J. Lightwave Technol. 2006; 24(23): 4711-4728.

18. Oxenløwe LK, Ji H, Galili M, Pu MH, Hu H, Mulvad, HCH, Yvind K, Hvam JM, Clausen AT, Jeppesen P. Silicon photonics for signal processing of Tbit/s serial data signals. IEEE J. Sel. Top. Quantum Electron. 2012; 18(2): 996-1005.

19. Pelusi MD, Ta'eed VG, Fu LB, Mägi E, Lamont MRE, Madden S, Choi DY, Bulla DAP, Luther-Davies B, Eggleton BJ. Applications of highly-nonlinear chalcogenide glass devices tailored for high-speed all-optical signal processing. IEEE J. Sel. Top. Quantum Electron. 2008; 14(3): 529-539.

20. Chowdhury A, Hagness SC, McCaughan L. Simultaneous optical wavelength interchange with a two-dimensional second-order nonlinear photonic crystal. Opt. Lett. 2000; 25(11): 832-834.

21. Chowdhury A, Staus C, Boland BF, Kuech TF, McCaughan L. Experimental demonstration of 1535–1555 nm simultaneous optical wavelength interchange with a nonlinear photonic crystal. Opt. Lett. 2001; 26(17): 1353-1355.

22. Wang J, Sun QZ. Theoretical analysis of power swapping in quadratic nonlinear medium. Appl. Phys. Lett. 2010; 96(8): 081108.

23. Wang J, Nuccio SR, Wu XX, Yilmaz OF, Zhang L, Fazal I, Yang JY, Yue Y, Willner AE. 40-Gbit/s optical data exchange between WDM channels using second-order nonlinearities in PPLN waveguides. NLO 2009, paper PDPA1, 2009.

24. Wang J, Nuccio SR, Wu XX, Yilmaz OF, Zhang L, Fazal I, Yang JY, Yue Y, Willner AE. 40 Gbit/s optical data exchange between wavelength-division-multiplexed channels using a periodically poled lithium niobate waveguide. Opt. Lett. 2010; 35(7): 1067-1069.

25. Wang J, Bakhtiari Z, Xiao-Li Y, Yilmaz OF, Nuccio SR, Wu XX, Huang H, Yang JY, Yue Y, Fazal I, Hellwarth R, Willner AE. Experimental demonstration of data traffic grooming of a single 10-Gbit/s TDM tributary channel between two 160-Gbit/s WDM channels. OFC 2010, paper OWF1, 2010.

26. Wang J, Bakhtiari Z, Yilmaz OF, Nuccio SR, Wu XX, Willner AE. 10 Gbit/s tributary channel exchange of 160 Gbit/s signals using periodically poled lithium niobate. Opt. Lett. 2011; 36(5): 630-632.

27. Mori K, Takara H, Saruwatari M. Wavelength interchange with an optical parametric loop mirror. Electron. Lett. 1997; 33(6): 520-522.

28. Gao Y, Dai YH, Shu C, He SL. Wavelength interchange of phase-shift-keying signal. IEEE Photon. Technol. Lett. 2010; 22(11): 838-840.

29. Wong KKY, Marhic ME, Uesaka K, Kazovsky LG. Demonstration of wavelength exchange in a highly nonlinear fiber. ECOC 2001, pp. 272-273, 2001.

30. Uesaka K, Wong KKY, Marhic ME, Kazovsky LG. Polarization-insensitive wavelength exchange in highly-nonlinear dispersion-shifted fiber. OFC2002, paper ThY3, 2002.

31. Uesaka K, Wong KKY, Marhic ME, Kazovsky LG. Wavelength exchange in a highly nonlinear dispersion-shifted fiber: theory and experiments. IEEE J. Sel. Topics Quantum Electron. 2002; 8(3): 560-568.

32. Fung RWL, Cheung HKY, Wong KKY. Widely tunable wavelength exchange in anomalous-dispersion regime. IEEE Photon. Technol. Lett. 2007; 19(22): 1846-1848.

33. Cheung HKY, Fung RWL, Kwok CH, Wong KKY. All-optical packet switching by pulsed-pump wavelength exchange in a highly nonlinear dispersion-shifted fiber. OFC2007, paper OTuB4, 2007.

34. Shen M, Xu X, Yuk TI, Wong KKY. Byte-level parametric wavelength exchange for narrow pulsewidth return-to-zero signals. IEEE Photon. Technol. Lett. 2009; 21(21): 1591-1593.

35. Kwok CH, Kuo BPP, Wong KKY. Pulsed pump wavelength exchange for high speed signal de-multiplexing. Opt. Express 2008; 16(15): 10894-10899.

36. Shen M, Xu X, Yuk TI, Wong KKY. A 160-Gb/s OTDM demultiplexer based on parametric wavelength exchange. IEEE J. Quantum Electron. 2009; 45(11): 1309-1316.

37. Shen M, Cheung HKY, Fung RWL, Wong KKY. A comprehensive study on the dynamic range of wavelength exchange and its impact on exchanged signal performance. J. Lightwave Technol. 2009; 27(14): 2707-2716.

38. Wang J, Bakhtiari Z, Xiao-Li Y, Nuccio SR, Yilmaz OF, Wu XX, Yang JY, Yue Y, Fazal I, Hellwarth R, Willner AE. Phase-transparent optical data exchange of 40-Gbit/s DPSK signals using four-wave-mixing in a highly nonlinear fiber. OFC 2010, paper OMT6, 2010.

39. Wang J, Bakhtiari Z, Nuccio SR, Yilmaz OF, Wu X, Willner AE. Phase-transparent optical data exchange of 40 Gbit/s differential phase-shift keying signals. Opt. Lett. 2010; 35(17): 2979-2981.

40. Wang J, Nuccio SR, Huang H, Wang X, Yilmaz OF, Wu XX, Yang JY, Yue Y, Willner AE. Demonstration of 100-Gbit/s DQPSK data exchange between two different wavelength channels using parametric depletion in a highly nonlinear fiber. ECOC 2010, paper Mo.1.A.4, 2010.

41. Wang J, Nuccio SR, Huang H, Wang X, Yang JY, Willner AE. Optical data exchange of 100-Gbit/s DQPSK signals. Opt. Express 2010; 18(23): 23740-23745.

42. Wang J, Huang H, Wang X, Yang JY, Willner AE. Optical phase-transparent data grooming exchange of multi-channel 100-Gbit/s RZ-DQPSK signals. IEEE 23rd Photonics Society Annual Meeting 2010, paper WN2, 2010.

43. Wang J, Huang H, Wang X, Yang JY, Willner AE. Multi-channel 100-Gbit/s DQPSK data exchange using bidirectional degenerate four-wave mixing. Opt. Express 2011; 19(4): 3332-3338.

44. Wang J, Huang H, Wang X, Yang JY, Yilmaz OF, Wu XX, Nuccio SR, Willner AE. 2.3-Tbit/s (23X100-Gbit/s) RZ-DQPSK grooming switch

(simultaneous add/drop, data exchange and equalization) using double-pass LCoS and bidirectional HNLF. OFC 2011, paper OTuE2, 2011.

45. Wang J, Huang H, Wang X, Yang JY, Willner AE. Reconfigurable 2.3-Tbit/s DQPSK simultaneous add/drop, data exchange and equalization using double-pass LCoS and bidirectional HNLF. Opt. Express 2011; 19(19): 18246-18252.

46. Christen L, Yilmaz OF, Nuccio SR, Wu XX, Fazal I, Willner AE. Tunable time-slot-interchange of 40-Gb/s optical packets using conversion/dispersion-based tunable 100-ns delays. OFC2008, paper OThA4, 2008.

47. Yilmaz OF, Christen L, Wu XX, Nuccio SR, Fazal I, Willner AE. Time-slot interchange of 40 Gbit/s variable length optical packets using conversion-dispersion-based tunable delays. Opt. Lett. 2008; 33(17): 1954-1956.

48. Wang J, Yilmaz OF, Nuccio SR, Wu XX, Bakhtiari Z, Xiao-Li Y, Yang JY, Huang H, Yue Y, Fazal I, Hellwarth R, Willner AE. Data traffic grooming/exchange of a single 10-Gbit/s TDM tributary channel between two pol-muxed 80-Gbit/s DPSK channels. CLEO 2010, paper CFJ5, 2010.

49. Wang J, Yilmaz OF, Nuccio SR, Wu XX, Willner AE. Orthogonal tributary channel exchange of 160-Gbit/s pol-muxed DPSK signal. Opt. Express 2010; 18(16): 16995-17008.

50. Suzuki J, Taira K, Fukuchi Y, Ozeki Y, Tanemura T, Kikuchi K. All-optical time-division add-drop multiplexer using optical fibre Kerr shutter. Electron. Lett. 2004; 40(7): 445-446.

51. Wang J, Willner AE. Review of robust data exchange using optical nonlinearities. International Journal of Optics 2012; 2012: Article ID 575429 doi: 10.1155/2012/575429.

52. Wang J, Yang JY, Fazal IM, Ahmed N, Yan Y, Willner AE, Dolinar S, Tur M. Experimental demonstration of 100-Gbit/s DQPSK data

exchange between orbital-angular-momentum modes. OFC2012, paper OW1I.5, 2012.

53. Wang J, Yang JY, Fazal IM, Ahmed N, Yan Y, Huang H, Ren YX, Yue Y, Dolinar S, Tur M, Willner AE. Terabit free-space data transmission employing orbital angular momentum multiplexing. Nature Photonics 2012; 6(7): 488-496.

54. Willner AE, Yilmaz OF, Wang J, Wu XX, Bogoni A, Zhang L, Nuccio SR. Optically efficient nonlinear signal processing. IEEE J. Sel. Topics Quantum Electron. 2011; 17(2): 320-332.

55. Tian Y, Xiao XS, Gao SM, Yang CX. All-optical switch based on two-pump four-wave mixing in fibers without a frequency shift. Appl. Opt. 2007; 46(23): 5588-5592.

56. Parameswaran KR, Fujimura M, Chou MH, Fejer MM. Low-power all-optical gate based on sum frequency mixing in APE waveguides in PPLN. IEEE Photon. Technol. Lett. 2000; 12(6): 654-656.

57. Wang J, Sun JQ, Sun QZ. Experimental observation of a 1.5 μm band wavelength conversion and logic NOT gate at 40 Gbit/s based on sum-frequency generation. Opt. Lett. 2006; 31(11): 1711-1713.

58. Wang J, Sun JQ, Sun QZ. Single-PPLN-based simultaneous half-adder, half-subtracter, and OR logic gate: proposal and simulation. Opt. Express 2007; 15(4): 1690-1699.

59. Allen L, Beijersbergen MW, Spreeuw RJC, Woerdman JP. Orbital angular momentum of light and the transformation of Laguerre–Gaussian laser modes. Phys. Rev. A 1992; 45(11): 8185-8189.

60. Franke-Arnold S, Allen L, Padgett M. Advances in optical angular momentum. Laser Photon. Rev. 2008; 2(4): 299-313.

61. Yao AM, Padgett MJ. Orbital angular momentum: origins, behavior and applications. Adv. Opt. Photon. 2011; 3(2): 161-204.

62. Gibson G, Courtial J, Padgett M, Vasnetsov M, Pas'ko V, Barnett S, Franke-Arnold S. Free-space information transfer using light beams carrying orbital angular momentum. Opt. Express 2004; 12(22): 5448-5456.

63. Djordjevic IB, Arabaci M, Xu L, Wang T. Spatial-domain-based multidimensional modulation for multi-Tb/s serial optical transmission. Opt. Express 2011; 19(7): 6845-6857.

64. Lerber TV, Honkanen S, Tervonen A, Ludvigsen H, Küppers F. Optical clock recovery methods: Review (Invited). Opt. Fiber Technol. 2009; 15(4): 363-372.

65. Mulvad HCH, Tangdiongga E, Waardt H, Dorren HJS. 40 GHz clock recovery from 640 Gbit/s OTDM signal using SOA based phase comparator. Electron. Lett. 2008; 44(2): 146-147.

66. Oxenløwe LK, Gómez-Agis F, Ware C, Kurimura S, Mulvad HCH, Galili M, Nakajima H, Ichikawa J, Erasme D, Clausen AT, Jeppesen P. 640-Gbit/s data transmission and clock recovery using an ultrafast periodically poled lithium niobate device. J. Lightwave Technol. 2009; 27(3): 205-213.

67. Takara H, Ono H, Abe Y, Masuda H, Takenaga K, Matsuo S, Kubota H, Shibahara K, Kobayashi T, Miaymoto Y. 1000-km 7-core fiber transmission of 10X96-Gb/s PDM-16QAM using Raman amplification with 6.5 W per fiber. Opt. Express 2012; 20(9): 10100-10105.

68. Liu X, Chandrasekhar S, Chen X, Winzer PJ, Pan Y, Taunay TF, Zhu B, Fishteyn M, Yan MF, Fini JM, Monberg EM, Dimarcello FV. 1.12-Tb/s 32-QAM-OFDM superchannel with 8.6-b/s/Hz intrachannel spectral efficiency and space-division multiplexed transmission with 60-b/s/Hz aggregate spectral efficiency. Opt. Express 2011; 19(26): B958-B964.

69. Al Amin A, Li A, Chen S, Gao G, Shieh W. Dual-LP11 mode 4X4 MIMO-OFDM transmission over a two-mode fiber. Opt. Express 2011; 19(17): 16672-16679.

70. Ryf R, Randel S, Gnauck AH, Bolle C, Sierra A, Mumtaz S, Esmaeelpour M, Burrows EC, Essiambre RJ, Winzer PJ, Peckham DW, McCurdy AH, Lingle R. Mode-division multiplexing over 96 km of few-mode fiber using coherent 6 × 6 MIMO processing. J. Lightwave Technol. 2012; 30(4): 521-531.

71. Koizumi Y, Toyoda K, Yoshida M, Nakazawa M. 1024 QAM (60 Gbit/s) single-carrier coherent optical transmission over 150 km. Opt. Express 2012; 20(11): 12508-12514.

CHAPTER 2

An orthogonal wavelet division multiple-access processor architecture for LTE-advanced wireless/radio-over-fiber systems over heterogeneous networks

Chinmaya Mahapatra1*, Victor CM Leung1 and Thanos Stouraitis2

1 Department of Electrical and Computer Engineering, University of British Columbia

2 Department of Electrical and Computer Engineering Department, University of Patras

ABSTRACT

The increase in internet traffic, number of users, and availability of mobile devices poses a challenge to wireless technologies. In long-term evolution (LTE) advanced system, heterogeneous networks (HetNet) using centralized coordinated multipoint (CoMP) transmitting radio over optical fibers (LTE A-ROF) have provided a feasible way of satisfying user demands. In this paper, an orthogonal wavelet division multiple-access (OWDMA) processor architecture is proposed, which is shown to be better suited to LTE advanced systems as compared to orthogonal frequency division multiple access (OFDMA) as in LTE systems 3GPP rel.8 (3GPP, http://www.3gpp.org/DynaReport/36300.htm). ROF systems are a viable alternative to satisfy large data demands; hence, the performance in ROF systems is also evaluated. To validate the architecture, the circuit is designed and synthesized on a Xilinx vertex-6 field-programmable gate array (FPGA). The synthesis results show that the circuit performs with a clock period as short as 7.036 ns (i.e., a maximum clock frequency of 142.13 MHz) for transform size of 512. A pipelined version of the architecture reduces the power consumption by approximately 89%. We compare our architecture with similar available architectures for resource utilization and timing and provide performance comparison with OFDMA systems for various quality metrics of communication systems. The OWDMA architecture is found to perform better than OFDMA for bit error rate (BER) performance versus signal-to-noise ratio (SNR) in wireless channel as well as ROF media. It also gives higher throughput and mitigates the bad effect of peak-to-average-power ratio (PAPR).

KEYWORDS

Heterogeneous networks (HetNet); Coordinated multipoint (CoMP); LTE advanced radio over fiber (LTE A-ROF); Orthogonal wavelength division multiple-access (OWDMA); processor; Orthogonal frequency division multiple access (OFDMA); Xilinx vertex 6 FPGA Bit error rate (BER) Signal-to-noise ratio (SNR); Peak-to-average-power ratio (PAPR)

1 INTRODUCTION

The diversity of applications used over the internet has resulted in a demand for increased speed (data rate) over the network and a need for accommodating more users per unit area. This demand has urged research communities to provide greener and more cost-efficient networks. Several research studies have been conducted over the last decade, proposing cost-efficient broadband architectures. Today, next-generation long-term evolution (LTE) systems using radio signals over optical fibers are evolving towards centralized architectures, as a promising solution to meet the ever-increasing demand for high-speed wireless connectivity. Centralized architectures, epitomized by micro base stations, femto and picocell base-station/access-point architectures, and mesh networking solutions have promised to provide several benefits, including reduced power consumption, enhanced radio spectrum utilization capacity, and diversity of next-generation wireless communication networks [1].

As radio spectrum is expensive and band-limited, centralized LTE advanced-radio over fiber (ROF) has attracted significant research interest. It focuses on the optimum construction and utilization of the hardware resources to cater an area of high traffic. A typical design uses optical fiber to move analog or digitized radiofrequency (RF) between the central facility and the remote sites [2]. Choosing optical fiber over conventional coaxial cables enables the usage of the enormous bandwidth provided by the fiber as well as almost error-free transmission for short ranges in a metro area network (MAN). Software-defined radio (SDR) provides efficient, cost-effective and easy-to-handle deployment architecture for the LTE A-ROF system. It follows a normal server/multi-client IT network and provides flexibility in architecture deployment. It also provides big savings of operational and infrastructure cost for service providers.

In the current LTE and Wi-Fi systems, orthogonal frequency division multiple-access (OFDMA) is the technology of choice [3]. OFDMA uses inverse fast Fourier transform (IFFT) at the transmitter and fast Fourier transform (FFT) at the receiver and allocates fixed resources to users for a given set of operating parameters. Despite its several advantages, if coupled with other components of the LTE A, the use of OFDMA increases the cost and utilization overhead of system resources. Moreover, it suffers from large implementation complexity, requiring a fixed allocation of resources to all the users, regardless of the present traffic as well as a high peak-to-average-power ratio (PAPR) [4].

Orthogonal wavelet division multiple access (OWDMA) has been proposed as a viable alternative to OFDMA in communication systems. Previous work concentrated on digital video broadcast, and results were only plotted for the BPSK modulation scheme [5, 6]. Raajan et al. [7] provided bit error rate (BER) performance graphs for all the wavelets and modulation schemes, but no hardware architecture was provided for the proposed system. Similarly, Tao et al. [8] and Liew et al. [9] analyzed orthogonal wavelet division multiplexing (OWDM) for signaling over wideband linear time-varying channels (LTV) but, again, did not provide any architecture for deployment. 1-D orthogonal wavelets have been used [10–14] for image processing applications.

The paper is organized in sections, where Section 3 provides a brief description of previous work, the definition of wavelet transform, and reasons for choosing 9/7 Daubechies lifting scheme for evolving the architecture. Section 4 describes the proposed OWDMA processor architecture and explains the different building blocks. In Section 5, pipelining is introduced in the architecture to reduce power consumption. Section 6 presents the synthesis and comparison results of resources and timing with other similar architectures. Section 7 gives the performance comparison for OWDMA and OFDMA, based on quality of constraint (QOS) metrics for LTE A-ROF systems. Finally, conclusions are offered in Section 8, followed by references.

2 KEY CONTRIBUTIONS

A sequential output-based parallel processing (SBPP) architecture for OWDM was proposed and evaluated for BER and PAPR [15]. Its deployment in LTE A future 3GPP rel.10 and above requires that its structure should be flexible enough to adapt according to channel conditions to different values of transform size in order to service uniformly the same number of users. The structure needs to accommodate both forward and inverse operations through a common control input. The architecture should be power efficient, easily controllable through a single control and should have input-output ports matching with other system sub-blocks that will satisfy the timing requirements of the whole system. Moreover, it is important for it to offer improved performance in terms of spectral efficiency (throughput), quality of service (better BER at the same signal-to-noise ratio (SNR)) and should fit well in radio-over-fiber systems. In this paper, an OWDMA architecture is developed that has significantly better performance, is easy to deploy, and consumes fewer resources than similar architecture available in the literature.

Analyzing the approaches described in [5–9] gives insight about the extensive research performed on the various solution approaches to problems of LTE OFDMA systems and provides proof that orthogonal wavelets are a better and viable alternative to the existing wireless systems. Although the analysis and evaluations were done for BER and PAPR, it lacks in a unified system implementation, resource analysis, and thorough performance evaluation for current LTE systems. Our major contribution in this paper is to deal with these shortcomings in the present knowledge and present an overall system level solution. Moreover, we also provide performance analysis for optical fiber medium.

3 ORTHOGONAL WAVELET DIVISION MULTIPLEXING

Fast fluctuations in the time domain or frequency-specific information in the time domain can only be revealed through a time/frequency analysis. The wavelet transform maps a time function into a two-dimensional function of 'a' (the scale) and 'τ' (the translation) of the Wavelet function along the time axis [16, 17]. The continuous waveform transform (CWT) of a signal $s(t)$ has been defined as

$$\mathrm{CWT}(a,\tau) = \frac{1}{\sqrt{a}} \int s(t)\psi \frac{(t-\tau)}{a} dt, \tag{1}$$

where t is the time, $\psi(t)$ is the basic (or mother wavelet), and $\psi((t-\tau)/a)$ is the translated baby wavelet [6] created by either stretching or compressing the mother wavelet.

3.1 Formulation of OWDM from the 9/7 filter using lifting

From the CWT, it is possible to construct the discrete wavelet transform (DWT) and the inverse DWT) from banks of matched high-pass filters (HPF) and low-pass filter (LPF) [18]. Single carrier systems tend to have high bit rates but low frequency resolution, whereas OFDM has many sublevels, each transferring at a low bit rate. Since the wavelet transform contains both time and frequency information, it is possible to effectively send different data rates in different sublevels, according to channel conditions. When considering the DWT, there are a number of mother wavelet families that need to be evaluated. To replace OFDM systems in a multipath environment having carrier and symbol interference, the wavelets need to be orthogonal and periodical. Also, the realization using discrete structures is important for purpose of implementation. Therefore, only three families of wavelet satisfy all the abovementioned constraints: Daubechies, Symlet, and Coiflet [19].

Figure 1 presents a comparison of signal-to-noise ratio vs. bit error rates for the Symlet 1, Coiflet 2, and Daubechies 2 similar-order filters [19]. It can be seen that the least resilient wavelet family is the Symlet, followed by the Coiflet, then the Daubechies, which appears to be better suited for implementation.

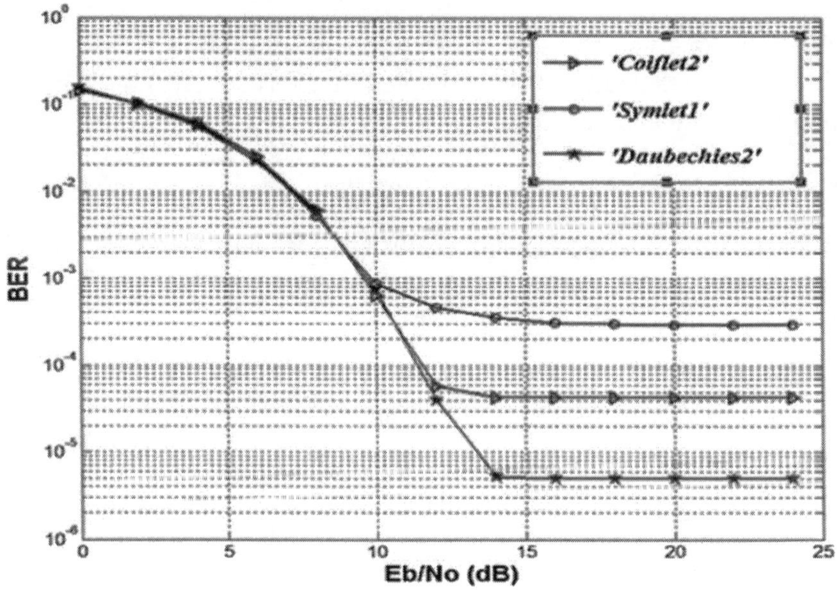

Figure 1. Signal-to-noise ratio versus bit error rate comparison for various orthogonal wavelet families.

The lifting scheme is used for the development of the architecture for a 9/7 Daubechies 1-D wavelet filter with two stages of lifting ($N = 2$), i.e., predict1 and update1, followed by predict2 and update2 in a second stage, followed by scaling [20, 21]. The basic idea of the lifting scheme is first to compute a trivial wavelet (or lazy wavelet transform) by splitting the original 1-D signal into odd- and even-indexed subsequences and then modify their values using alternating prediction and updating steps [22, 23]. The lifting algorithm consists of the following three steps:

- Split step

The original signal, $X(n)$, is split into odd and even samples (lazy wavelet transform).

- Lifting step

This step is executed as N sub-steps (depending on the type of the filter), where the odd and even samples are alternatingly filtered by the prediction and update filters.

- Scaling step

After N lifting steps, scaling coefficients K and $1/K$ are applied, respectively, on the odd and even samples, in order to obtain the low-pass band and the high-pass sub-band.

Orthogonal wavelet division multiple access (OWDMA) is a system, in which the wavelet domain is used to separate the sub-band components in the same way as OFDMA. The big difference between OFDMA and OWDMA is that in OFDMA, the FFT performs sub-band decomposition with a specific number of sub-bands at well-defined intervals, while OWDMA may dynamically allocate the number of sub-bands and the bandwidth of each [24].

4 OWDMA PROCESSOR ARCHITECTURE FOR LTE A AND LTE A-ROF

From the SBPP-OWDM scheme presented in the previous section, it is found that the final scaling and dilation coefficients are interdependent on predict and update outputs at each stage; thus, there is a delay and it also affects throughput. The structure requires two update and predict blocks to be implemented. OWDMA scheme requires that the structure should be flexible enough to adapt to different values of N, according to the channel conditions. The structure needs to accommodate both forward and inverse operations through a common control. The multiplicative coefficients for the filter need to be stored in a hardware-friendly format which will reduce the number of multiplication operations. Thus, a new OWDMA processor architecture has been developed that caters to all the requirements of a multiple-access system mentioned above. Moreover, parallelism is exploited in the architecture, along with pipelining, to formulate an efficient, low-power, and resource-friendly processor.

Predict and update block1 and predict and update block2 are combined together along with the scaling. In the forward operation, the wavelet coefficients for odd and even samples are calculated using (2) and (3). The odd values are calculated using a structure implementing a 9-tap finite impulse response (FIR) filter, and the even values are found out with a 7-tap FIR filter structure, as shown in Figure 2. Odd index wavelet coefficients first, third, $n-3$, and $n-1$, and even index wavelet coefficients second, $n-2$, and n th have adjusted input values fed using the symmetry property of the filter. The values of k before the first value 'X [1]' and after the last value '$X[N]$' are replaced by their symmetric values by using $x[k] = x[k+2i]$, where i takes the value $-k+1$ w.r.t to the input index k at the left side of the axis, and it takes the value $N-k$ at the right side of the axis. The FO $_C(j)$ and FE $_C(l)$ are forward odd and even filter coefficients, respectively, where $j = 1,2...9$ and $l = 1,2....7$.

$$\begin{aligned} Y_{\text{EVEN}}[N] = {} & \text{FE}_C(1) \times X[N-3] + \text{FE}_C(2) \times X[N-2] \\ & + \text{FE}_C(3) \times X[N-1] + \text{FE}_C(4) \times X[N] + \\ & + \text{FE}_C(5) \times X[N+1] + \text{FE}_C(6) \times X[N+2] \\ & + \text{FE}_C(7) \times X[N+3]; \{N = \text{EVEN}\} \end{aligned} \quad (2)$$

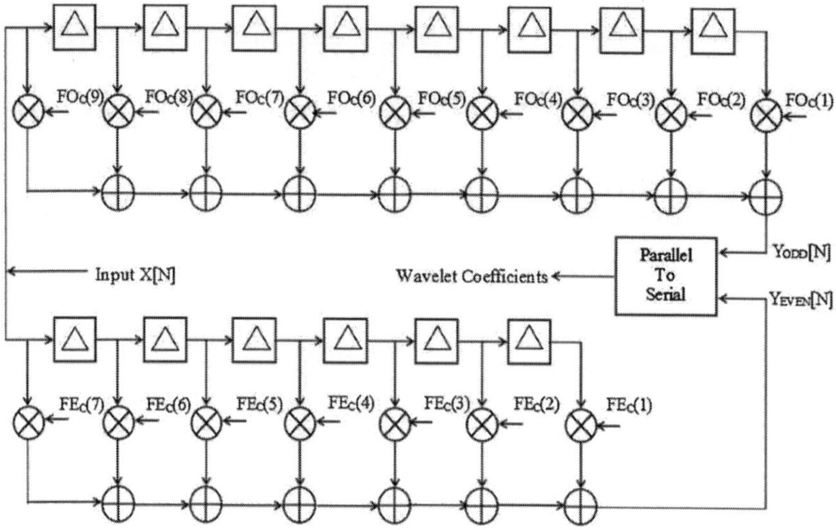

Figure 2. The core filter unit showing 9-tap and 7-tap FIR filter structures with input X **[** N **]**. Δ represents a delay of 1 unit, + represents an addition operation, and X represents a multiplication operation.

$$\begin{aligned}Y_{\mathrm{ODD}}[N] = {} & \mathrm{FO}_C(1) \times X[N-4] + \mathrm{FO}_C(2) \times X[N-3] \\ & + \mathrm{FO}_C(3) \times X[N-2] + + \mathrm{FO}_C(4) \times X[N-1] \\ & + \mathrm{FO}_C(5) \times X[N] + \mathrm{FO}_C(6) \times X[N+1] + \\ & + \mathrm{FO}_C(7) \times X[N+2] + \mathrm{FO}_C(8) \times X[N+3] \\ & + \mathrm{FO}_C(9) \times X[N+4]; \{N = \mathrm{ODD}\}\end{aligned} \tag{3}$$

In the inverse operation, the wavelet coefficients for odd and even samples are calculated using (4) and (5). The odd values are calculated using a structure implementing a 7-tap FIR filter, and the even values are found with a 9-tap FIR filter structure. Odd index wavelet coefficients first, third, and $n-1$, and even value wavelet coefficients second, fourth, $n-2$, and n th have adjusted input values from symmetry in a similar way as described in the previous paragraph for the forward operation. The boundary conditions are formulated using a state machine control logic implementation elaborated in the control unit section. The IO $_C$ (l) and IE $_C$ (j) are inverse odd and even filter coefficients, respectively, where j = 1,2...9 and l = 1,2....7.

$$\begin{aligned} X_{\mathrm{EVEN}}[N] = {} & \mathrm{IE}_C(1) \times Y[N-4] + \mathrm{IE}_C(2) \times Y[N-3] + \mathrm{IE}_C(3) \\ & \times Y[N-2] + +\mathrm{IE}_C(4) \times Y[N-1] + \mathrm{IE}_C(5) \\ & \times Y[N] + \mathrm{IE}_C(6) \times Y[N+1] + +\mathrm{IE}_C(7) \\ & \times Y[N+2] + \mathrm{IE}_C(8) \times Y[N+3] + \mathrm{IE}_C(9) \\ & \times Y[N+4]; \{N = \mathrm{EVEN}\} \end{aligned} \tag{4}$$

$$\begin{aligned} X_{\mathrm{ODD}}[N] = {} & \mathrm{IO}_C(1) \times Y[N-3] + \mathrm{IO}_C(2) \times Y[N-2] + \mathrm{IO}_C(3) \\ & \times Y[N-1] + \mathrm{IO}_C(4) \times Y[N] + +\mathrm{IO}_C(5) \\ & \times Y[N+1] + \mathrm{IO}_C(6) \times Y[N+2] + \mathrm{IO}_C(7) \\ & \times Y[N+3]; \{N = \mathrm{ODD}\} \end{aligned} \tag{5}$$

The proposed OWDMA processor consists of a core unit to multiply filter coefficients with delayed input, accumulate with previous values, and compute the wavelet coefficients. Its control unit controls which coefficients are to be applied at the complex multiplier input. A coefficient generator unit reads the appropriate coefficients from memory. The OWDMA unit acts as slave to a master scheduler unit that feeds it with clock, address, input data, and variables. Figure 3 shows the high-level architecture of the OWDMA processor unit with the scheduler. The architecture is a two-parallel structure due to simultaneous calculation of odd and even data. The scheduler and the three major units of the proposed system, namely, the core unit, control unit, and the coefficient generator unit, are discussed below.

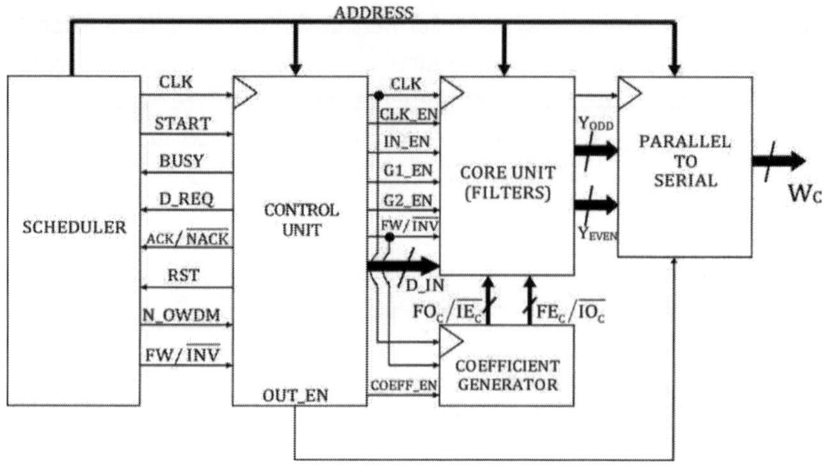

Figure 3. The top-level generic OWDMA processor implementation block diagram.

4.1 Scheduler

The proposed OWDMA processor can be interfaced with the scheduler, according to the scheme presented Figure 3. In this scheme, the scheduler communicates with the OWDMA processor using a set of dedicated hand-shaking signals. The scheduler acts as the master, sets the address of the processor, and provides clock to it (CLK). First, the scheduler requests the control unit block to initiate a new transform using the START signal. The controller unit sets the BUSY signal low, if it is ready to start the process for the new transform, or high, if it is in the middle of an already continuing process. When the controller is ready, it sends a data request (D_REQ) signal to the scheduler, which then responds with the input data. If the controller correctly gets the input, it sends an acknowledgement (ACK) signal; otherwise, it sends NACK NACK$\overline{\text{NACK}}$, and the scheduler retransmits. Along with the data input, it sends the information for the size of OWDM (N_OWDM) as well as the forward/inverse operation (FW/INV FW/INV$\text{FW}/\overline{\text{INV}}$) signal. The OWDMA processor uses the RST signal to indicate the end of data, when it completes the transform. At the same time, it sets the BUSY signal low to indicate to the scheduler that it is ready to start a new transform.

4.2 Core unit

The core unit consists of two FIR filter units. One is 9-tap and the other is a 7-tap, as shown in Figure 2. They both have CLK, CLK_EN, IN_EN, G1_EN, G2_EN, D_IN, and FW/INV ------ FW/INV-FW/$\overline{\text{INV}}$, as common inputs, and Y_{ODD} and Y_{EVEN} as the odd and even filter outputs. The only difference is the inputted coefficients for the multiply and accumulate units inside the FIR filters. The FO $_C$(1...9), IE $_C$(1...9) are the coefficient inputs for the 9-tap filter block, and FE $_C$(1...7), IO $_C$(1...7) are coefficient inputs for the 7-tap filter block. These coefficients are explained in more detail in Section 4.4. The CLK input counts from 0 to $N + 4$ and then gets reset. The extra 5 clock cycles after the normal N-cycles are for flushing the output to 0. D_IN is the data input. IN_EN is enable signal for data input. G1_EN and G2_EN are enable signals for switches that switch input gates and are enabled by the control logic. FW/INV ------ FW/INV-FW/$\overline{\text{INV}}$ signal is for forward (a '0') and inverse operation (a '1'). The outputs of both filters are fed to a parallel-to-serial converter block that downsamples the data and rearranges the coefficients to give the final coefficients W_C. It has an OUT_EN (output enable) signal to start calculating the output W_C (the wavelet coefficients). Figure 4 shows all the logic signals in a timing diagram.

4.3 Control unit

The control unit consists of two separate logic units for forward and inverse computation and is implemented using a finite state machine having five states: S0, S1, S2, S3, and S4. It toggles on the positive CLK edge input, and at each state, the output controls IN_EN, G1_EN, G2_EN,

OUT_EN, FW/INV------ INV$\overline{INV}$, _COEF_EN (0/1), and FW/INV------ FW/INV$FW/\overline{INV}$. The FW/INV------ FW/INV$FW/\overline{INV}$ signal controls which control the logic unit is to be used (forward or inverse). G1_EN and G2_EN are gate control switches that switch inputs for the delay registers at the boundary conditions.

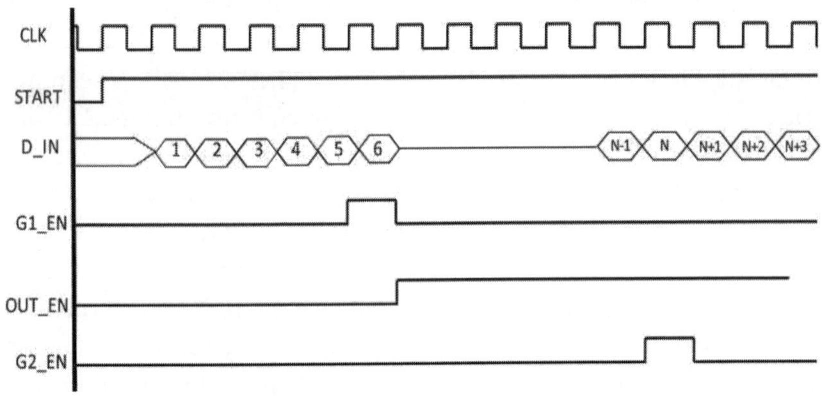

Figure 4. Timing diagram showing logical signals with clock.

The input value has to be symmetrically extended at the boundaries to avoid distortion. '*X* [1]' is the first input value and no previous value is available. Using the symmetric property of the lifting scheme [20] as shown in Figure 5, the next input value '*X* [2]' is extended to the left of '*X* [1]' and is used to perform the filter operation. Similarly, at the end of the row, the input value '*X*[N]' is the last one. By copying the input value '*X*[N-1]' to the right of '*X*[N]', the boundary condition at the right end can be satisfied. The control logic is shown in Figure 6. At the positive edge count value of 4, G1_EN is enabled for 1 clock cycle. G2_EN is enabled at n th clock value. Output is calculated starting from fifth clock count to $(n + 4)$th count and then everything resets back to state S0.

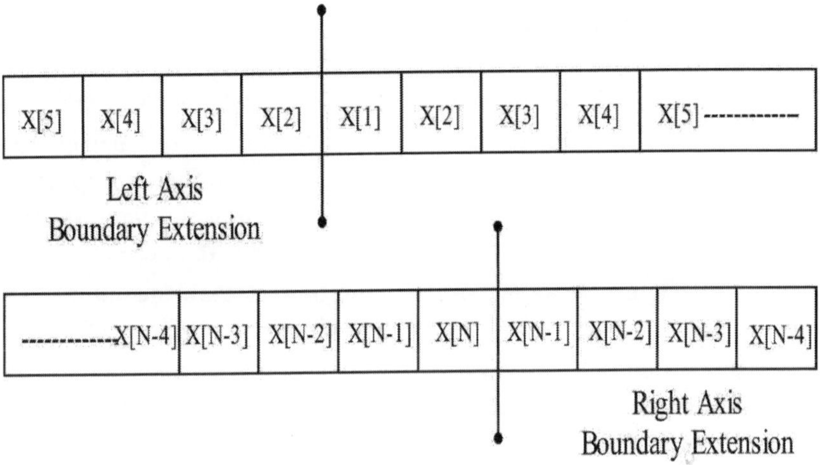

Figure 5. Symmetric boundary extension of input data.

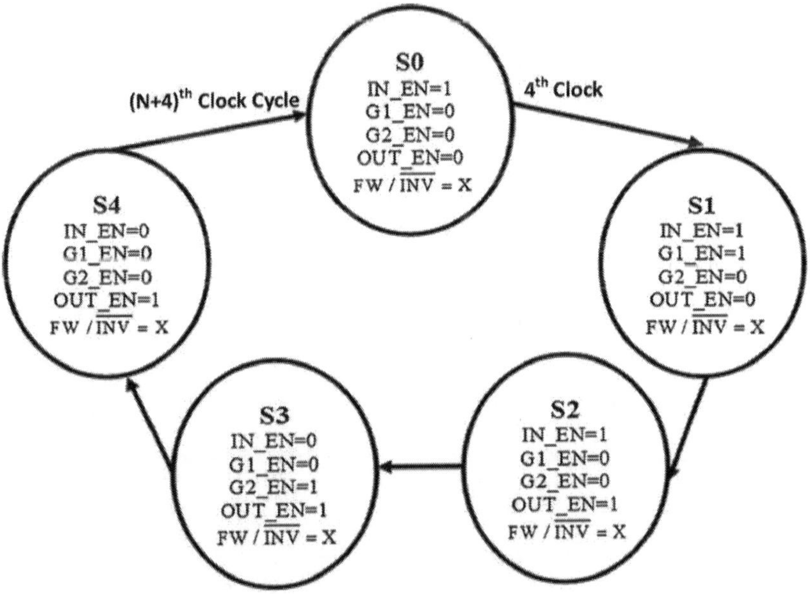

Figure 6. Control logic finite state machine implementation.

4.4 Coefficient generator unit

The coefficient generator block is a memory block that contains the odd and even filter coefficients to be multiplied during forward/inverse operation. Providing the appropriate constant to the multiplier, it implements the desired multiplication. The width of the multipliers is determined by the accuracy of the constants and the data path bitwidth. The drawback of the above implementation is that the multipliers occupy a great amount of area and restrict the throughput of the processing unit. Using shift-add operations to replace the multiplications with constants optimizes the above implementation and results in an improved processing block.

To perform shift and add operations, coefficients are converted in two's complement Q.15 format. That is, they are shifted 15 bits to the left and converted to their respective two's complement binary value. The filter constants are quantized, taking in account the number of bits with value '1', in their positive representation. That is because each '1' yields a term to be summed. For example, the sets of odd and even coefficients for the forward path are shown in Tables 1 and 2, respectively, similarly, the coefficients for the reverse path can be defined.

Table 1. Forward odd coefficients

Index	Q.15 format	Binary format
$FO_c(1)$	−1,240	1111101100101000
$FO_c(2)$	781	0000001100001101
$FO_c(3)$	9,956	0010011011100100
$FO_c(4)$	−16,358	1100000000011010
$FO_c(5)$	30,031	0111010101001111
$FO_c(6)$	−16,358	1100000000011010
$FO_c(7)$	9,956	0010011011100100
$FO_c(8)$	781	0000001100001101
$FO_c(9)$	−1,240	1111101100101000

5 PIPELINING THE PARALLEL ARCHITECTURE

Power dissipation is a major drawback of the system in the downlink and especially the uplink of a LTE A network. In the proposed OWDMA processor architecture, pipelining the stages of the 9-tap and 7-tap filters, along with the two-stage parallel structure, can help save in power budget. The saved power can be used to accommodate more number of users or increase the range of the system. Pipelining reduces the effective critical path by introducing latches along the critical data path. The critical path (or the minimum time required for processing a new sample) is limited by 1 multiply and 8 add times in the 9-tap filter structure and 1 multiply and 6 add times in the 7-tap filter structure of the OWDMA processor, respectively, as shown in Figure 2. Thus, the 'sample period' is given by (6) and (7), where T_{sample} (9-tap) and T_{sample} (7-tap) are the sampling frequencies of the respective filters

$$T_{\text{sample}}(9\text{–tap}) \geq T_{\text{M}} + 8 \times T_{\text{A}} \tag{6}$$

$$T_{\text{sample}}(7\text{–tap}) \geq T_{\text{M}} + 6 \times T_{\text{A}} \tag{7}$$

Pipelining is accomplished by introducing 14 and 10 additional latches in the feed-forward path of the 9-tap and the 7-tap filter structure, respectively, thereby reducing the critical path to $T_{\text{M}} + T_{\text{A}}$ for both filters. In an M-level pipelined system, the number of delay elements in any path from input to output is $(M - 1)$ greater than that in the same path in the original sequential circuit. Thus, we apply eight-level pipelining to the 9-tap filter circuit and six-level pipelining to the 7-tap filter circuit. When the sample speed does not need to be increased, this can be used for lowering the power consumption. The power dissipation (P_{CMOS}) in any circuit depends on the total capacitance C_{total} of the CMOS logic, the supply voltage V_{CC}, and the clock frequency f. The total power depends on static and dynamic power consumption:

$$P_{\text{CMOS}} = C_{\text{total}} \times V_{\text{CC}}^2 \times f, f = 1/T_{\text{seq}} \tag{8}$$

With increase in propagation/gate delay, leakage current decreases, thereby reducing the static power consumption of the system [25]. The propagation delay (T_{pd}) depends on the charging capacitance C_{charge} in a clock cycle and the difference $(V_{CC} - V_t)^2$, where V_t is the threshold voltage:

$$T_{\text{pd}} = \frac{C_{\text{charge}} \times V_{\text{CC}}}{k(V_{\text{CC}} - V_{\text{t}})^2} \tag{9}$$

Applying pipelining reduces the capacitance to be charged/discharged in one-clock period, while the inherent parallel processing allows for increasing the clock period for charging/discharging the original capacitance. In an L-parallel system ($L = 2$ in our case), the clock period of the circuit is increased to LT_{pd}. This implies that the supply voltage can be reduced to βV o $(0 < \beta < 1)$. Hence, the power consumption, compared with the original system, is reduced by a factor β^2. The propagation delay of the L-parallel, M-pipelined filter is obtained as

$$LT_{\text{pd}} = \frac{(C_{\text{charge}}/M) \times \beta V_{\text{CC}}}{k(\beta V_{\text{CC}} - V_{\text{t}})^2} = \frac{L \times C_{\text{charge}} \times V_{\text{CC}}}{k(V_{\text{CC}} - V_{\text{t}})^2} \tag{10}$$

Finally, we can obtain the following equations to compute β and power dissipated in OWDM circuit, respectively

$$\beta^2 \times V_{CC} \times L \times M - \beta \times \left[(V_{CC} - V_t)^2 + 2 \times V_{CC} \times V_t \times L \times M\right] + V_t^2 \times L \times M = 0 \quad (11)$$

$$P_{OWDM} = \beta^2 \times P_{CMOS} \quad (12)$$

6 PERFORMANCE RESULTS AND COMPARISONS

6.1 Synthesis of the proposed architecture and resource utilization

In order to evaluate the performance of the architecture, it is required to make use of certain metrics that characterize the architecture in terms of the hardware resources used and the computation time. The hardware resources used for filtering are measured by the number of multipliers an number of adders, while those used for the storage of data and filter coefficients are measured by the number of registers. In general, the computation time is technology dependent. However, a metric that is technology independent and can be used to determine the computation time (T) is the number of clock cycles (N_{CLK}) elapsed between the first and the last samples inputted to the architecture. Assuming that the clock period is T_c, the total computation time can then be obtained as $T = N_{CLK} \times T_c$.

To validate the circuit design based on the proposed architecture, the implementation is done on a test bed that includes one central processor with multiple distributed antenna nodes and multiple mobile stations. The test bed operates in the 2.4-GHz ISM band for its license-exempt convenience. The central processor consists of RF front ends with 20- to 80-MHz bandwidth, a number of 125 ~ 250-MHz 14-bit ADC/DACs mounted on the latest Xilinx Virtex-6 FPGA digital signal processing (DSP)/FPGA processing unit (San Jose, CA, USA) to pre-process data samples. All the carry propagation adders of the architecture have a 16-bit word length and use a structure that combines the carry-skip and carry-select adders [26]. The FPGAs inside the platforms are XC6VLX75T-2, which is capable of operating at a clock frequency of 650 MHz at supply voltage of $V_{CC} = 2.5$ V and quiescent voltage of $V_t = 1.5$ V. The resources utilized by the FPGA implementation in terms of the numbers of configuration logic block (CLB) slices, flip-flop slices, DSP 48's, input/output blocks (IOBs), and block RAMs (BRAMs) are given in Table 3.

Table 3. FPGA resource summary for OWDMA

Resource	Used	Percentage
CLB slices	959	6
Flip-flop slices	653	1
DSP 48's	32	11
IOBs	180	75
BRAMs	3	2

The implemented circuit is found to perform well with a clock period as short as 7.036 ns (i.e., a maximum clock frequency of 142.13 MHz) for a transform size of $N = 512$. By replacing the values of V_{CC}, V_t, L, and M in (18) and (19), it can be found out that the power consumption on the chip on which the circuit is implemented is reduced by a factor of (1/9). The new power usage is only 143 mW per antenna.

6.2 Comparison with other architectures

The proposed OWDMA architecture is compared next with various 1-D wavelet architectures, as well as with available commercial OFDMA chips. Computation time (T), CLB slices or area occupied on FPGA, maximum clock frequency and area/speed ratio are some of the key performance metrics that are compared. Table 4 gives the comparison results between different 1-D wavelet architectures present in the literature. For even comparison, $N = 512$ is taken as the size of the input data in all architectures. It can be inferred from the table that although our proposed architecture consumes a few more hardware resources as compared to recursive architecture [10], parallel FDWT [12], pipelined [13], and Arch1D-II [14], it significantly performs better in terms of maximum clock frequency and computation time. It is also seen that the area to speed ratio is the second lowest for our architecture. Although the parallel FDWT implementations [12] present a better area/speed ratio, high computation times make them unsuitable for high-speed applications.

Table 4. Comparison between various 1-D architectures

Architecture	*N* (size of computation)	No. of CLB slices	*F*max (MHz)	Time (µs)	Area/speed ratio	Filter type
Recursive architecture [7]	512	439	50	10.25	8.78	1-D (9/7)
Symmetrically extended [8]	512	1,279	44.1	8.36	29	1-D
Parallel FDWT [9]	512	850	171.8	2.98	4.95	1-D (9/7)
Pipelined [10]	512	785	85.49	6	9.18	1-D
Arch1D-II [11]	512	921	136	1.88	6.77	1-D (9/7)
Proposed OWDMA processor	512	959	142.13	1.82	6.75	1-D (9/7)

Current advanced 4G systems deploy OFDMA architecture. So, it becomes imperative to compare our proposed OWDMA processor architecture with the state-of-the-art implementation. The OFDMA core uses the Radix-4 and Radix-2 decompositions for computing the discrete Fourier transform (DFT) [27, 28]. When using the Radix-4 decomposition, the N-point FFT consists of log4 (N) stages, with each stage containing $N/4$ Radix-4 butterflies. Point sizes that are not a power of 4 need an extra Radix-2 stage for combining data. An N-point FFT using Radix-2 decomposition has log2 (N) stages, with each stage containing $N/2$ Radix-2 butterflies. The comparison between the two architectures is depicted in Table 5. It can be seen that there is at least 90% improvement in the computation time when OWDMA core is used as opposed to OFDMA core for a 10% increase in DSP resources on the FPGA, whereas the total area occupied on FPGA remains comparable with the respective N-point computation. Furthermore, it has been shown [6] that the OWDMA which implements wavelet transforms has a complexity of $O(N)$, as opposed to the OFDMA that contains FFT operations and has complexity of $O(N \log_2 N)$.

Table 5. Comparison of resource utilization for OFDMA and OWDMA processors

Technique	Architecture	N (size of computation)	No. of CLB slices	No. of LUT slices	BRAM	DSP slices	Time (μs)
OFDMA processing	R2	256	842	650	3	3	4.18
	R2L	1,000	1,025	839	3	3	31.58
	R2		1,106	882	3	6	18.63
	R2L	2,000	1,137	882	5	3	66.74
	R2		1,082	952	5	6	39.41
Proposed OWDMA processing	Parallel and pipelined	256	614	355	3	32	0.91
	Parallel and pipelined	1,000	1,338	1,042	3	32	3.61
	Parallel and pipelined	2,000	1,160	848	5	32	7.26

7 QUALITY COMPARISON IN 4G LTE A/LTE A-ROF SYSTEM

In any communication system, BER versus SNR and spectral efficiency (throughput) are standard quality of service (QoS) parameters, which give a measure of system performance. Therefore, the proposed OWDMA architecture has been compared to existing OFDMA architectures in 4G LTE A systems with respect to the above two QoS parameters. Figure 7 shows the difference between the two implementations with respect to a practical system perspective. As can be seen in Figure 7a,b, the difference is in the IFFT/FFT blocks of OFDMA that are replaced by OWDM modulator/OWDM demodulator block. The number of samples N_{OFDM} is same as N_{OWDM} that is modulated and demodulated input/output in the respective paths. The inherent processing is similar for both OFDM and OWDM, the only difference being the way the samples are treated in the respective processors. So, there is no need to make any significant changes to the overall processing samples. Only replacing the IFFT/FFT blocks of OFDMA with that of OWDM modulator/OWDM demodulator will serve our purpose.

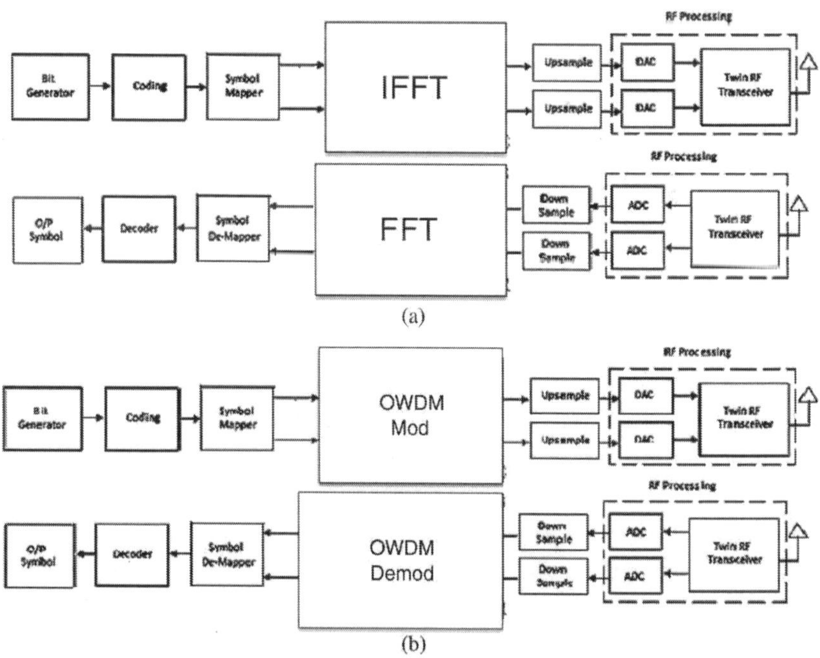

Figure 7. OFDM transmitter and receiver (a) and OWDM transmitter and receiver (b).

Figure 8a gives an overview of the proposed centralized CoMP system based on an existing LTE backbone with fiber connectivity. The system consists of a central processing unit (CPU) that contains the Xilinx FPGA and DSPs for all the data processing for uplink and downlink. There are four separate processing modules inside the CPU model in Figure 8a, each having transmit (*Tx*) and receive (*Rx*) data processing capability. The four modules are connected to a 4 × 4 hub that, in turn, is connected to a RF switch, capable of switching in *Tx* and *Rx* directions. The transmit and receive directions have an electrical-to-optical convertor (laser diode) to convert from electrical to optical signal to be transmitted via the fiber to a remote antenna unit (RAU), located at different places in a cell site. The laser diode is modulated by the RF signal in the downlink path. The resulting intensity-modulated optical signal is then transmitted through the single-mode fiber towards a photodiode. The received optical signal is converted to RF signal (optical-to-electrical convertor) by direct detection through a PIN photodetector. The signal is then amplified and radiated by the antenna. The optical fibers cover in practice a distance of few hundred meters to a few kilometers, enough to cover a building or a small area. The RAU is a passive unit containing only optical to electrical convertor and amplifier and a RF antenna at the 2.4/5-GHz band to transmit or receive radio signals. The RAUs are relatively close to the user equipment, generally within few hundred meters. So, an International Telecommunication Union (ITU) pedestrian multipath channel with Doppler frequency $f_d = 5$ Hz is chosen for simulations.

Figure 8b,c shows the transmitter and receiver units, respectively, of the proposed architecture. In the first step of the transmitter processing, the user data are generated, depending on the previous acknowledgement (ACK) signal. If the previous user data transport block (TB) was not acknowledged, the stored TB is retransmitted using a hybrid automatic repeat request (HARQ) scheme. Then, a cyclic redundancy check (CRC) is calculated and appended to each user's TB. The data of each user are independently encoded using a turbo encoder with quadrature permutation polynomial (QPP)-based interleaving [29]. Each block of coded bits is then interleaved and rate-matched with a target rate, depending on the received channel quality indicator (CQI) user feedback. The encoding process is followed by data modulation, which maps the channel-encoded TB to complex modulation symbols. Depending on the CQI, a modulation scheme is selected for the corresponding resource block. Modulation schemes used for downlink-shared channel (DL-SCH) here are QPSK, 16-QAM, and 64-QAM. The modulated transmit symbols are then mapped to a multiple-input and multiple-output (MIMO) precoding matrix. The optimum precoding matrix is selected from a code book, depending on the pre-coding control information (PCI) that is fed back from the user equipment (UE) to the transmitter. Finally, the individual symbols to be transmitted are mapped to the resource elements. Downlink reference symbols and synchronization symbols are also inserted into the OFDM/OWDM time-frequency grid. The assignment of a set of resource blocks (RBs) to UEs is carried out by the scheduler based on the CQI reports from the UEs.The receiver structure is shown in Figure 8c. Each UE receives the signal transmitted by the evolved node B (eNB) and performs the reverse physical-layer processing of the transmitter. First, the receiver has to identify the RBs that carry its designated information. The estimation of the channel is performed using the reference signals available in the resource grid. Based on this channel estimation, the quality of the channel may be evaluated, and the appropriate feedback information calculated. The channel knowledge is also used for the demodulation and soft decoding of the OFDM/OWDM signal. In case of MIMO, a MMSE decoder is used. Finally, the UE performs HARQ combining and channel decoding. In order to cut down the processing time after the end of every turbo iteration, a CRC check of the decoded block is performed and, if correct, decoding is stopped.

The path from UE to RAU, i.e., the uplink uses single carrier-frequency division multiple access (SC-FDMA) as OFDM, has high PAPR. High PAPR requires expensive and inefficient power amplifiers with high requirements on linearity, which increases the cost of the user terminal and also drains the battery faster. Since OWDM has better PAPR, it can be used in the uplink as well.

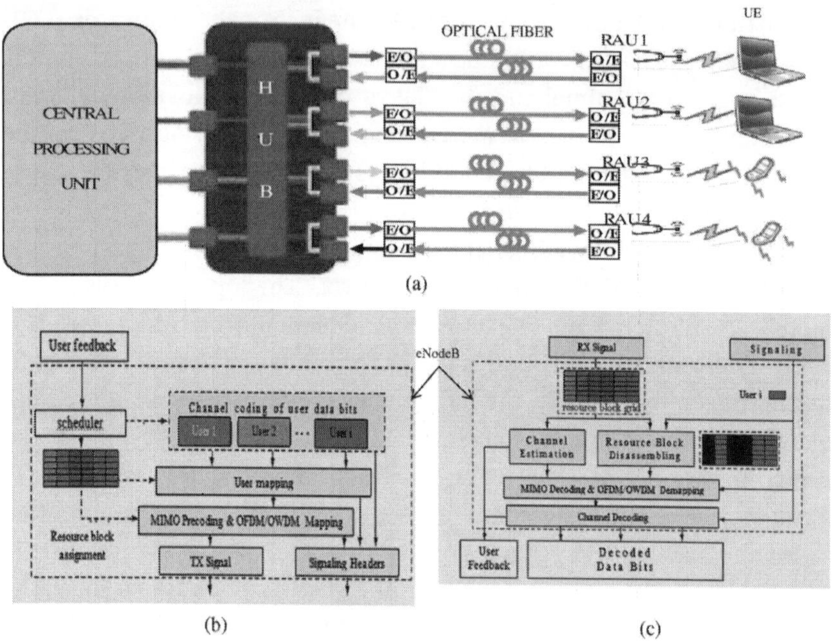

Figure 8. Deployment diagram of 4G LTE A-ROF systems having centralized architecture. **(a)** Overview of centralized eNB architecture. **(b)** Transmitter unit. **(c)** Receiver unit.

7.1 Bit error rate comparison

Using the above system model, standard QoS parameters BER and throughput have been compared for OFDM and OWDM architectures. In addition, a performance evaluation of radio over single-mode fiber system using coded OFDM and OWDM and the relation of fiber length with BER is discussed. A comparison is also drawn between peak-to-average-power ratio in the two systems.

Figure 9a,b,c shows the simulation results for bit-error-rate versus signal-to-noise ratio for OWDMA/OFDMA systems for different modulation formats. Table 6 shows the parameters used for the simulations. The plots are drawn for modulation schemes of QPSK, 16-

QAM, and 64-QAM for an ITU-defined extended pedestrian A model (EPA) with Doppler frequency $f_d = 5$ Hz. The size of transform for both the architectures is taken as 1,024, for a total of 50 resource blocks at 20-MHz bandwidth. Comparisons are drawn taking both uncoded and rate-1/2 turbo-coded OWDMA/OFDMA systems. We can infer from the graphs that OWDMA performs better than OFDMA for both coded and uncoded schemes. At QPSK modulation, as in Figure 9a, OWDMA shows a performance improvement close to 1.5 dB, which increases to nearly 2.5 dB for 64-QAM modulation, as shown in Figure 9c. The gain in SNR can be used to either increase reach or accommodate more users.

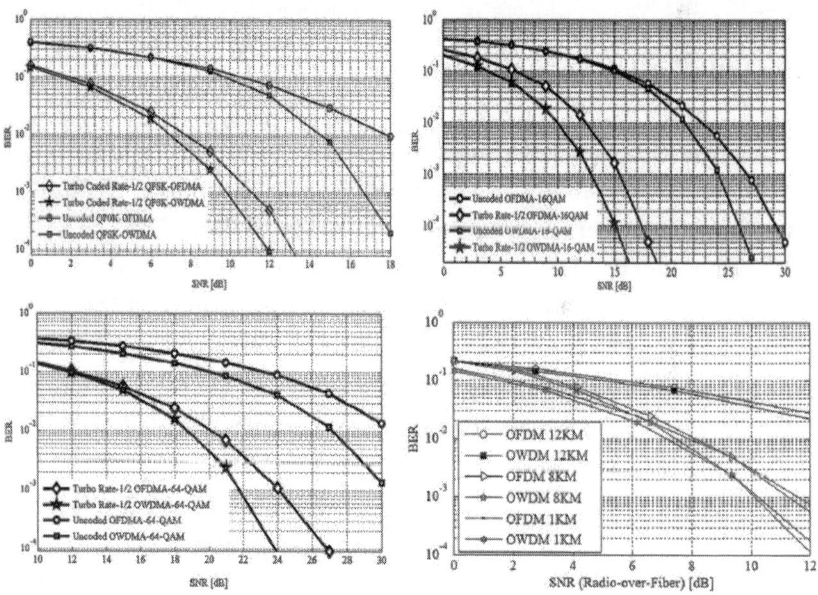

Figure 9. Signal-to-noise ratio vs. bit error rate comparison for OWDMA and OFDMA architectures in LTE A-ROF systems. (a) Signal-to-noise ratio vs. bit error rate comparison for rate-1/2 turbo-coded and uncoded QPSK-OWDMA and QPSK-OFDMA. **(b)** Signal-to-noise ratio vs. bit error rate comparison for rate-1/2 turbo-coded and uncoded 16-QAM-OWDMA and 16-QAM-OFDMA. **(c)** Signal-to-noise ratio vs. bit error rate comparison for rate-1/2 turbo-coded and uncoded 64-QAM-OWDMA and 64-QAM-OFDMA. **(d)** Signal-to-noise ratio (radio-over-fiber) vs. bit error rate comparison for QPSK-OWDMA and QPSK-OFDMA at different fiber lengths.

Table 6. Simulation parameters.

Parameter	Value
Channel bandwidth	20 MHz
Modulation schemes	QPSK, 16-QAM, 64-QAM
Multiple access architectures	OWDMA/OFDMA
N (size of computation)	1,024
Error control code	Rate-1/2 turbo codes
Channel model	ITU extended pedestrian A model (EPA) with $f_d = 5$ Hz

The relation of optical fiber length with BER in terms of performance (SNR of radio-over-fiber) is shown in Figure 9d. Table 7 defines the simulation parameters for the optical link. The SNR of radio-over-fiber is defined as the ratio of probe power (P_{FIBER}) of the fiber and the total noise in the ROF system. Total noise of the ROF system is the sum of additive white Gaussian noise (AWGN) plus the fiber nonlinearities and it passes through an ITU multipath fading channel (EPA channel), as defined in the previous case. The SNR of ROF is given by (13), where h is the radio channel coefficient, and N_{AWGN} and $N_{FIBER\text{-}NL}$ are the AWGN noise and degradation from fiber nonlinearities, respectively.

$$\mathrm{SNR}_{ROF} = \frac{|h|^2 \times P_{FIBER}}{N_{AWGN} + |h|^2 \times N_{FIBER\text{-}NL}} \tag{13}$$

We find that there is no significant degradation on the BER performance until the fiber length becomes 12 km, due to considerable modal dispersion. OWDMA shows slightly better performance as compared to the OFDMA system. This shows that the OWDM-ROF system ensures high service availability over long distances up to 8 km, which came in accordance with standard distances between the indoor (baseband) and outdoor (radio) units.

Table 7. Fiber parameters in optical link

Parameter	Value
Type of fiber	Single-mode NZDSF
Fiber wavelength	1.55 μm
Fiber dispersion	4 ps/nm/km
Fiber attenuation	0.21 dB/km
Frequency spacing	50 GHz
Optical probe power	15 dBm
Span length	1, 8, and 12 km

7.2 Throughput of the OWDMA system

In this section, we compare the throughput of the OWDMA systems as compared to that of the OFDMA systems. In a fading system, the capacity of a massive MIMO system is given by [30]

$$C = B \times F \times \log_2(1 + M \times \beta \times \text{SNR}) \tag{14}$$

Here, SNR is the signal-to-noise ratio, B the bandwidth occupied by the data subcarriers, β defines the combined path loss and lognormal shadowing at any RAU, M is the number of RAUs, and F is a correction factor. The transmission of an OFDM signal requires also the transmission of a cyclic prefix (CP) to avoid inter-symbol interference and the reference symbols for channel estimation. The correction factor F represents the loss due to cyclic prefix. In order to avoid any ISI, the symbol time should be greater than the channel delay spread. The substreams should also be orthogonal to each other, and thus, OFDM is used. Assume the transmission of an N-sample sequence, $x[n] = \{x[0], x[1], \ldots, x[N-1]\}$, through a channel with L multipaths, $h[l] = \{h[0], h[1], \ldots, h[L-1]\}$. We assume that the channel consists of L distinct and resolvable paths, and $v[n]$ is assumed to be. So, the discrete sampled received signal, $r[n]$, at the output of the channel can be written as

$$r[n] = \sum_{l=0}^{L-1} h[l] \times x[n] + v[n] \tag{15}$$

It is well known that multiplication in the DFT domain corresponds to the circular convolution in the time domain. In order to achieve circular convolution using linear convolution, we must add a prefix that is the 'cyclic prefix' onto the transmitted signal. This cyclic prefix makes the linear convolution appear as a circular convolution and represents a loss in the achievable data rate that becomes significant in the highly fading channels. But in the case of OWDM that uses wavelet transform, the operations involve shift and multiply operations with filter coefficients. The shift by two for subsequent pairs of rows produces a downsampling operation within the matrix transformation and also makes the matrix orthogonal and circulant [31]. Therefore, a cyclic prefix is not required in the case of OWDM. This gives a significant throughput advantage particularly in highly dispersive channels.

Figure 10 shows the throughput (Mb/s) versus SNR in decibels with values of M (RAUs) ranging from 1 to 4 for OWDM and OFDM systems. Shannon's limit for the capacity at different M is also drawn to show the upper bound. The red circles depict the throughput for OWDM/OFDM at a particular modulation scheme and SNR. It is found that the OWDM system achieves better throughput at same channel conditions putting less burden on the overall resources in terms of modulation used, at a given SNR. For instance, OWDM with QPSK provides 25% better throughput performance as compared to OFDM with 16-QAM between 12- and 14-dB SNR. Data rates up to 760 Mb/s can be achieved using OWDM systems, and the results are very close to Shannon's limit. Thus, OWDM is not only a more efficient system than OFDM but a system that performs close to Shannon's limit.

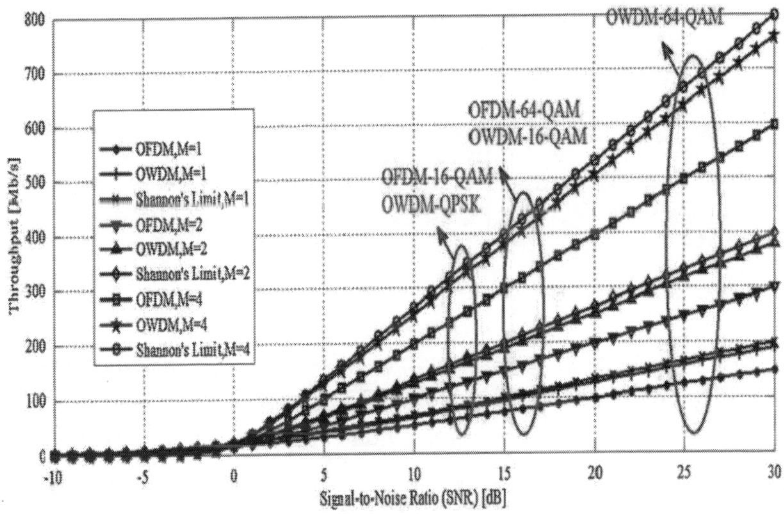

Figure 10 . Signal-to-noise ratio versus throughput (spectral efficiency) for OWDMA and OFDMA systems at different values of M .

7.3 Peak-average-to-power ratio

PAPR appears to be a main disadvantage of OFDM. In this paper, we show that PAPR performance in OWDMA is better than that of the OFDMA systems. PAPR depends on the bandwidth efficiency of the system. The problem of high PAPR is usually associated with OFDM because it is much easier to reach high bandwidth efficiency for OFDM. There exist a large number of publications in literature devoted to the PAPR problem in OFDM, mostly claiming only a slight improvement on the exiting architecture. The results of tests indicate that OWDM is the ideal candidate to solve the existing PAPR issues and still deliver high bandwidth efficiency. Our PAPR simulation and analysis is carried out based on its complementary cumulative distribution (CCD). So, for a given $PAPR_0$ (dB), the percentage of combinations that guarantee ($PAPR > PAPR_0$) is a meaningful criterion for analysis. In general, the PAPR of wavelet coefficients coming out is defined in (16) as the ratio between the maximum instantaneous power and its average power, where $E[W_C(t)]^2$ is the average power of $W_C(t)$ (wavelet coefficients).

$$\mathrm{PAPR(dB)} = 10 \times \log_{10}\left[\frac{\max\left|\left|W_C(t)\right|^2\right|}{E\left|W_C(t)\right|^2}\right] \tag{16}$$

The PAPR performance metric we consider is the complementary cumulative distribution function (CCDF), which is plotted as in Figure 11. The OWDMA system provides reduced PAPR of around 2 dB, which is favorable for RF amplifier operation as compared to OFDMA systems.

8 CONCLUSION

In this paper, we have developed a flexible, hardware-friendly, and low-power OWDMA architecture design for deployment in ROF systems having LTE-advanced configuration. The key contribution of the paper is the architecture derived for a LTE A-ROF system with an interface of input and output ports that can replace the OFDMA block offering added benefits.

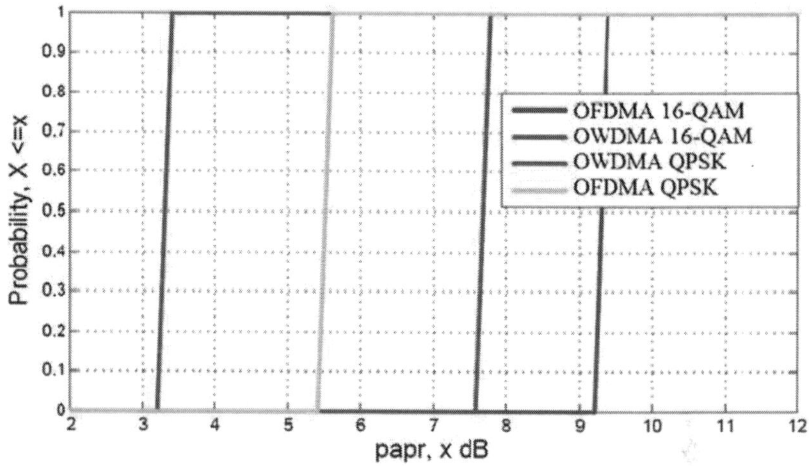

Figure 11. CCDF plot of PAPR in decibels.

We first derived an architecture based on previous 9/7 lifting scheme wavelet filters. The computation of the method is described using filters, controller, and parallel-to-serial units. The scheduler is also implemented for easy interfacing of the sub-block with other blocks of the system. The architecture is validated on a centralized processor having Xilinx Virtex-6 FPGAs at $N = 512$. We compare our architecture with various other 1-D 9/7 wavelets available in the literature as well with existing OFDMA implementations. We also compute the quality parameters BER, throughput, and PAPR for OWDMA and compare them with the existing OFDMA systems.

We found that our architecture runs at a speed of 142.13 MHz, consuming only 143 mW of power per antenna. It is better, in terms of resource consumption, as compared to other similar 1-D 9/7 implementations. We also found that it is also significantly better than OFDMA systems in terms of resource utilization and BER, throughput, and PAPR performance for ROF systems. Hence, it is shown that the OWDMA systems are well suited for high data rate communications and also can accommodate more users.

ACKNOWLEDGEMENTS

Research performed and documented in this thesis was supported by the Canadian Natural Sciences and Engineering Research Council (NSERC) through grant STPGP 396756.

REFERENCES

1. Haoming L, Hajipour J, Attar A, Leung VCM: Efficient HetNet implementation using broadband wireless access with fiber-connected massively distributed antennas architecture. Wirel. Comm. IEEE. 2011, 18(3):72-78.

2. Attar A, Haoming L, Leung VCM: Green last mile: how fiber connected massively distributed antenna systems can save energy. Wirel. Comm. IEEE 2011, 18(5):66-74.

3. 3GPP: Technical specification group radio access network; (EUTRA) and (E-UTRAN); overall description; stage 2. 2008. . Accessed 26 Nov 2013 http://www.3gpp.org/DynaReport/36300.htm

4. Jiang T, Imai Y: An overview: peak-to-average power ratio reduction techniques for OFDM signals. IEEE Trans. Wirel. Comm. 2008, 57: 56-57.

5. Linfoot SL: Wavelet families for orthogonal wavelet division multiplex. Electron. Lett. 2008, 44(18):1101-1102. 10.1049/el:20081681

6. Linfoot SL, Ibrahim MK, Al-Akaidi MM: Orthogonal wavelet division multiplex: an alternative to OFDM. Consum. Electron. IEEE Trans. 2007, 53(2):278-284.

7. Raajan NR, Monisha B, Kumar MR, Philomina AJ, Priya MV, Parthiban D, Suganya S: Design and implementation of orthogonal wavelet division multiplexing (OHWDM) with minimum bit error

rate. Paper presented at the 3rd international conference on trends in information sciences and computing (TISC), Chennai; 2011:122-127.

8. Tao X, Leus G, Mitra U: Orthogonal wavelet division multiplexing for wideband time-varying channels. Paper presented at the IEEE international conference on acoustics, speech and signal processing (ICASSP), Prague; 2011:3556-3559.

9. Liew BA, Berber SM, Sandhu GS: Performance of a multiple access orthogonal wavelet division multiplexing system. Volume 2. Paper presented at the third international conference on information technology and applications (ICITA), Sydney; 2005:350-353.

10. Liao HY, Mandal MK, Cockburn BF: Efficient architectures for 1-D and 2-D lifting-based wavelet transforms. IEEE Trans. Signal Process. 2004, 52(5):1315-1326. 10.1109/TSP.2004.826175

11. McCanny P, Masud S, McCanny J: Design and implementation of the symmetrically extended 2-D wavelet transform. ICASSP 2002, 3: 3108-3111.

12. Raghunath S, Aziz SM: High speed area efficient multi-resolution 2-D 9/7 filter DWT processor. Paper presented at the IFIP international conference on very large scale integration, Nice; 2006:210-215.

13. Masud S, McCanny J: Reusable silicon IP cores for discrete wavelet transform applications. IEEE Trans. Circuits Syst. I, Reg. Papers1 2004, 51(6):1114-1124. 10.1109/TCSI.2004.829236

14. Uzun IS, Amira A: Rapid prototyping—framework for FPGA based discrete biorthogonal wavelet transforms implementation. IEEE Vision Image Signal Process 2006, 153(6):721-734. 10.1049/ip-vis:20045080

15. Mahapatra C, Ramakrishnan A, Stouraitis T, Leung VCM: A novel implementation of sequential output based parallel processing - orthogonal wavelet division multiplexing for DAS on SDR platform.

Paper presented at the 19th IEEE international conference on electronics, circuits and systems (ICECS), Seville; 2012:320-323.

16. Chan YT: Wavelet Basics. Kluwer Academic Publishers, Dordrecht; 1994.

17. Daubechies I: Ten Lectures on Wavelets. 3rd edition. SIAM, Philadelphia; 1994.

18. Mallat SG: A theory for multiresolution signal decomposition: the wavelet representation. IEEE Trans. Pattern Anal. Machine Intell. 1989, 11(7):674-693. 10.1109/34.192463

19. Linfoot SL: A study of different wavelets in orthogonal wavelet division multiplex for DVB-T. IEEE Trans. Consum. Electron. 2008, 54(3):1042-1047.

20. Jamin A, Mahonen P: Wavelet packet modulation for wireless communications. J. Wirel. Commun. Mob. Comput. 2005, 5(2):123-137. 10.1002/wcm.201

21. Cheng C, Parhi KK: High-speed VLSI implementation of 2-D discrete wavelet transform. Signal Process. IEEE Trans. 2008, 56(1):393-403.

22. Sweldens W, Daubechies I: Factoring wavelet transforms into lifting steps. J. Fourier Anal. Appl. 1998, 4: 247-270. 10.1007/BF02476026

23. Sweldens W: Lifting scheme: a new philosophy in biorthogonal wavelet constructions. In Proceedings of the SPIE Conference on Wavelet Application in Signal and Image Processing III. Volume 2569. Edited by: Laine AF, Unser M. SPIE, Bellingham; 1995:68-79. 10.1117/12.217619

24. Akansu N, Medley MJ: Wavelet and subband transforms: fundamentals and communication application. IEEE Commun. Mag. 1997, 35: 104-115.

25. Qi W, Vrudhula SBK: An investigation of power delay trade-offs for dual Vt CMOS circuits. Paper presented at the international conference on computer design (ICCD), Austin; 1999:556-562.

26. Zhang C, Wang C, Ahmad MO: A pipelined VLSI architecture for high-speed computation of the 1-D discrete wavelet transform. IEEE Trans. Circuits Syst. I, Reg. Papers1 2010, 57(10):2729-2740.

27. Lin YW, Lee CY: Design of an FFT/IFFT processor for MIMO OFDM systems. IEEE Trans. Circuits Syst. I, Reg. Papers1 2007, 54(4):807-815.

28. Liu H, Lee H: A high performance four-parallel 128/64-point radix-24 FFT/IFFT processor for MIMO-OFDM systems. Paper presented at the IEEE Asia Pacific conference on circuits and systems, Macao; 2008:834-837.

29. 3GPP: Technical specification group radio access network; evolved universal terrestrial radio access (E-UTRA); multiplexing and channel coding (release 8). 2008. . Accessed 26 Nov 2013 http://www.3gpp.org/ftp/Specs/archive/36_series/36.212/

30. Mahboob S, Mahapatra C, Leung VCM: Energy-Efficient Multiuser MIMO Downlink Transmissions in Massively Distributed Antenna Systems with Predefined Capacity Constraints. Paper presented at the seventh international conference on broadband, wireless computing, communication and applications (BWCCA), Victoria, Canada; 2012:208-211.

31. Dilmaghani R, Ghavami M: Comparison between wavelet-based and Fourier-based multicarrier UWB systems. Commun. IET 2008, 2(2):353-358. 10.1049/iet-com:20070181

CHAPTER 3

Mixed-Signal Architectures for High-Efficiency and Low-Distortion Digital Audio Processing and Power Amplification

Sergio Saponara and Pierangelo Terreni

Department of Information Engineering, University of Pisa

ABSTRACT

The paper addresses the algorithmic and architectural design of digital input power audio amplifiers. A modelling platform, based on a meet-in-the-middle approach between top-down and bottom-up design strategies, allows a fast but still accurate exploration of the mixed-signal design space. Different amplifier architectures are configured and compared to find optimal trade-offs among different cost-functions: low distortion, high efficiency, low circuit complexity and low sensitivity to parameter changes. A novel amplifier architecture is derived; its prototype implements digital processing IP macrocells (oversampler, interpolating filter, PWM cross-point deriver, noise shaper, multilevel PWM modulator, dead time compensator) on a single low-complexity FPGA while off-chip components are used only for the power output stage (LC filter and power MOS bridge); no heatsink is required. The resulting digital input amplifier features a power efficiency higher than 90% and a total harmonic distortion down to 0.13% at power levels of tens of Watts. Discussions towards the full-silicon integration of the mixed-signal amplifier in embedded devices, using BCD technology and targeting power levels of few Watts, are also reported.

1. INTRODUCTION

Small size, low-cost and high-efficiency audio amplifiers, integrated as much as possible with digital audio signal processing tasks in the same embedded device, are required in several consumer applications: home and car entertainment, computer/portable multimedia players and, for low power levels, hearing aids devices. Conventional linear amplifiers feature low-distortion performance but have several disadvantages versus market needs [1, 2]: they are too heavy and energy inefficient and the achievable power density is limited by the physical size and cost of cooling hardware and power devices. An extra Digital-to-Analog Converter (DAC) is needed, before the analog amplifier, for digital sources: CD, Super Audio CD and DVD supports, MP3 files, and Digital Audio Broadcasting.

To achieve similar low-distortion performance of linear amplifiers but with a higher power efficiency, and hence smaller size and cost, the recent research has been focused on switching amplifiers. Several class D PWM (Pulse Width Modulation) topologies or hybrid class A–D or B–D ones have been proposed in literature [3–6]. However such topologies are still analog input amplifiers. An alternative solution is the direct amplification of the digital source based on the switching architecture at the bottom of Figure 1: in the digital domain the input PCM (Pulse Code

Modulation) signal is directly converted in a PWM one; the latter is amplified by an inverter power bridge, switching at hundreds of kHz, and provided to the speaker after lowpass LC filtering.

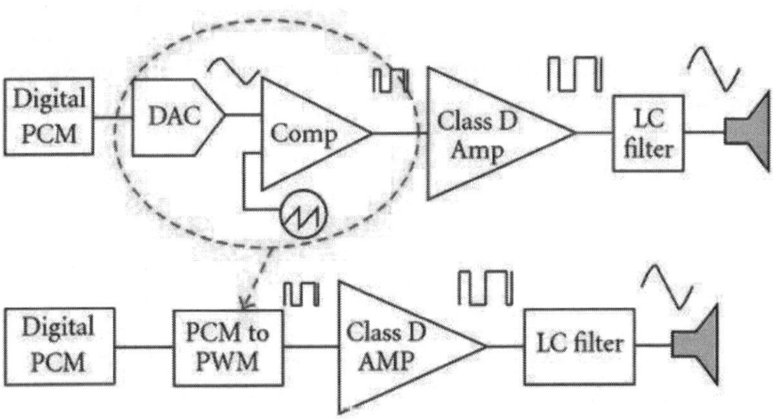

Figure 1. DAC plus class D amplifier versus digital power audio amplifier.

1.1. The State of the Art of Digital Power Audio Amplifiers

A direct conversion of the input PCM stream to PWM is not useful: PWM is a nonlinear technique and the intermodulation between the PWM carrier frequency and the baseband audio signal leads to poor-quality amplifiers [7]. To this aim, in academia and industry [8–24], several techniques have been proposed to improve the basic scheme in Figure 1: new digital audio processing algorithms [12–15, 17–20, 22–24] or novel feedback schemes [9–11, 21] or multilevel PWM power bridges [16] have been published to correct the distortions introduced by the PWM modulation or by the nonideal behaviour of the power stage. However the new proposed techniques require extra hardware resources; the performance gain is paid in terms of increased circuital complexity and cost. The overall amplifier often requires multiple chips for the digital part (including digital signal processor and/or ASIC and/or FPGA), plus ADC and high-order analog filters for the feedback plus multiple power transistors and gate drivers for multi level PWM. As a result high-performance systems require high circuit complexity, implemented using multiple chips and often multiple-boards, and are not suitable for consumer applications or embedded devices. Solutions with lower complexity are usually obtained at the expense of audio quality reduction.

1.2. Aim and Outline of This Work

This work explores the design space of digital audio amplifiers to find an optimal mixing of different analog and digital techniques. The resulting architecture aims at achieving optimal performance in terms of low-distortion and high power efficiency while still allowing a low-cost implementation: all the digital processing part integrated in a single device, for example, a low-complexity FPGA, plus off-chip components only for the power stage, made up of a MOS H-bridge and an LC filter but without any heat sink.

Most state-of-the-art techniques propose specific optimizations for just a part of the scheme in Figure 1; when integrating together different known techniques the relevant hardware overheads add up while the extra gain in performance can be negligible. However, the exhaustive design space exploration of digital power audio amplifiers is not straightforward since it needs fast but still accurate models involving the codesign of heterogeneous components: computation intensive processing algorithms at functional level with hardware components at physical level; low power digital and mixed-signal circuits with analog power devices; silicon integrated circuits with discrete devices. Hereafter Section 2 presents a platform-based modelling flow and the cost metrics used to drive the design space analysis. The models used to find optimal trade-offs between complexity, power efficiency, distortion and sensitivity are presented in Section 3 together with architectural comparison results. Section 4 shows the prototyping of the selected architecture targeting power levels of tens of Watts. Section 5 compares the obtained results versus the state of the art and discusses the extension of the work to fully integrated amplifiers for power levels of few Watts. Conclusions are drawn in Section 6.

2. PLATFORM-BASED DESIGN FLOW AND METRICS

2.1. Design Metrics Definition

The definition of the multiple cost metrics to be optimized is essential to drive the design space exploration and the correct comparison of different architectures. The design metrics are the audio signal distortion, the power efficiency, the circuit complexity and the architecture sensitivity to parameter changes. The input signals are PCM samples with a bit size *n* from 16 to 24 and a sample frequency F_{IN} between 44 kS/s and 96 kS/s (the lower values for frequency and bit-size are typical of audio CD while the higher values are used in audio DVD). The target output power, Pout, amounts to tens of Watts with power efficiency levels up to 90%. Reported data in this paper refer to the example case of max Pout of 70 Wrms (or 35 Wrms) delivered to a 4 Ω(or 8 Ω) speaker. The total harmonic distortion (THD) considered for High-Fidelity (Hi-Fi) is a level lower than 0.2%, the optimal target is 0.1%. As discussed in [1] there are high-end products for professional applications, using linear amplifiers, with THD figures below 0.001%; however the subjective sensitivity of the hearing human system to THD levels below 0.3% is often negligible. Most Hi-Fi amplifiers, for example, Sony STR-DE445 [13], for consumer home or car markets have a THD of 0.2%. The considered frequency response in this paper is 20 Hz–20 kHz, although THD optimizations focus on the range 500 Hz to 2 kHz where the hearing human system is mainly sensible and very often music signals are below 16-17 kHz [1]. The target circuit complexity for the digital processing circuitry amounts to tens of equivalent ASIC gates, a value that can be fitted in a single low-cost FPGA leaving space to integrate other audio processing tasks thus realizing a complete audio acquisition/playing system in a single embedded device.

2.2. Platform-Based Design Flow

To allow a fast but still accurate design space exploration we followed a meet-in-the-middle approach between bottom-up and top-down strategies [25]. A configurable modelling platform has been built starting from libraries of analog and digital building block components. A library of accurate spice models has been derived bottom-up for the hardware components whose nonideal characteristics and nonlinearity affect the behaviour of the power audio amplifier: power MOS, power supply, analog filters, OpAmp and comparators optionally used in the feedback loop. As example, we have created Spice models for the power MOS in [26], used also in the prototyping phase; such models consider all key electrical parameters [27] and their dependence on input driving signal (V_{GS}) and output delivered power: MOS transfer curves, drain source breakdown voltage BV_{DS}, R_{DSon} resistance, gate charge Q_G, body diode reverse recovery charge Q_{rr}, internal gate resistance R_G, MOS rise and fall times $T_r$and $T_f$and switching frequency F_{sw}, transistor packaging and thermal characteristics. The Spice models have been integrated with parametric and fixed-point Simulink models for the signal processing algorithms proposed to enhance the basic scheme in Figure 1, see details in Section 3. The resulting Spice/Simulink environment is then used top-down to build multiple architectures (proper configuring and combining the building block models) and to allow their fast but still accurate comparison. The considered design metrics are those in Section 2.1. This analysis allows a first selection of the most promising architectures; for them the comparison is further refined in a second step using HDL models for the digital audio algorithms.

Synthesized on different technologies (standard-cells CMOS libraries or SRAM-based FPGAs) the HDL models permit the evaluation of the gate complexity and power consumption of the digital circuitry. The selected architectures are finally prototyped, and the real performances measured, using FPGA technology plus a discrete power output stage.

3. MIXED-SIGNAL ARCHITECTURAL EXPLORATION

This section presents the modelling platform which includes the following building blocks, see Figure 2: an over sampler, a cross point estimator for natural PWM, a sigma-delta noise shaper, a multilevel PWM generator, a dead time insertion unit, a power bridge, an LC filtering stage, a feedback loop with PWM signal compensation. To each building block parametric functional and cost models have been associated. For some blocks multiple algorithms are implemented. By combining the different building blocks and configuring their parameters, different possible architectures have been obtained, simulated and compared. The following subsections detail the architectures and functionalities implemented in each block of Figure 2, the relevant parameters, and the results obtained from the comparison of different configurations. The most suited choices for block combination and parameter configuration are also highlighted.

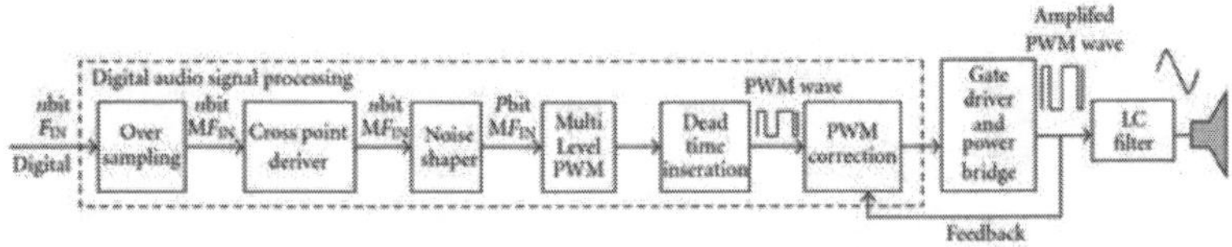

Figure 2. Modelling platform for digital input power audio amplifier.

3.1. Oversampler

To reduce the output THD an oversampler is added before the PCM to PWM conversion in the digital domain, see Figure 2. Oversampling by a parametric factor *M* is realized first inserting *M*-1 zeros after each original sample (zero padding); the data stream is then processed with an interpolating filter to remove high frequency spurious repetitions of the baseband signal. The higher the oversampling factor *M* is, the higher ($\uparrow M$) the PWM carrier frequency is and hence the transition bandwidth for the LC output filter used to remove intermodulation distortion. Thus, a high over-sampling factor *M* simplifies the design of the LC analog filter at the expense of an increased complexity for the digital interpolating filter.

During simulations to find the most suited sizing for the factor *M* and for the oversampling filter structure, the masks in Table 1 have been used to specify the filter magnitude response. Mask 1 is known in literature [28, 29] while mask 2 is a new proposal of this work. Compared to mask 1, mask 2 features more stringent requirements in terms of pass-band ripple and stop-band attenuation (needed to meet a target THD $<0.2\%$) but allows for a larger transition band; indeed music signals rarely exceed the 17 kHz imposed by a pass-band of 0.4 F_{IN} and hence the pass-band 0.45 F_{IN} in mask 1 is often excessive.

Table 1. Interpolating filter mask specifications.

Mas type	Pass-band	Stop-band	Pass-band ripple	Stop-band attenuation
1	0.45 F_{IN}	0.55 F_{IN}	<0.1 dB	>50 dB
2	0.4 F_{IN}	0.6 F_{IN}	<0.02 dB	>60 dB

Once specified the magnitude response of the filter it is important to define its architectural implementation (FIR or IIR, windowing type, direct or cascade multistage structure, parallel or iterative implementation of the multiply and accumulate-MAC- unit, data bit width), since its hardware cost can represent the main bottleneck of the whole audio system [28–32].

For the audio interpolating filter an FIR approach has to be preferred to an IIR for its linear phase response. However to implement the masks in Table 1 with an oversampling factor $M = 8$a direct implementation of an FIR type needs a number of taps, that is, a filter order *Z*, of some hundreds as reported in Table 2. Equiripple and Kaiser FIR types are considered in Table 2; analysis carried out using Gaussian, Hamming and Hanning FIR windows lead to similar results of the Kaiser FIR type. Such high filter orders entail a high computational complexity which amounts to roughly $Z \cdot M$MAC operations per sample. On the contrary using an IIR structure the same magnitude response can be obtained with a much lower filter order: see Table 2 which considers Butterworth, Chebychev and elliptic IIR filters.

Table 2. FIR and IIR order for the interpolating filter, oversampling x8.

	FIR		IIR		
	Equiripple	Kaiser	Butt.	Cheb.	Elliptic
Mask 1	193	233	38	13	7
Mask 2	129	146	24	11	7

The filter computational burden further increases when considering oversampling factors *M* much higher than 8. As example Table 3 reports the PWM carrier frequency $F_C = M \cdot F_{IN}$, the filter order and the MAC operations needed for Equiripple and Kaiser FIR filters when varying the oversampling factor from 8 to 128 with CD-quality $F_{IN} = 44.1$ kS/s.

From Table 3 it is clear that (i) the computational cost of a direct FIR implementation amounts to tens of millions MAC/s for $M = 8$and becomes prohibitive, in the range of tens of Giga MAC/s, when *M* grows up to 128; (ii) considering off-chip power output bridges (see Section

3.6) with typical PWM frequencies within 1 MHz, oversampling factors *M* higher than 8 or 16 should be avoided.

Table 3. Interpolating filter complexity and PWM frequency versus *M*.

M	Mas type	Equiripple		Kaiser		F_C
		order	$10^6 \cdot$ MAC/s	order	$10^6 \cdot$ MAC/s	kHz
8	1	193	68.1	233	82.2	352.8
8	2	129	45.5	146	51.5	
16	1	386	272.4	469	330.9	705.6
16	2	257	181.3	291	205.3	
32	1	772	1089	938	1324	1411.2
32	2	515	726,8	581	819,9	
64	1	1544	4358	1875	5292	2822.4
64	2	1029	2904	1161	3277	
128	1	3087	17425	3749	21162	5644.8
128	2	2057	11611	2321	13102	

To reduce the computational cost of FIR filters while keeping the advantages of their better phase response versus IIR structures, the following solutions can be adopted [29].

1 Polyphase filter implementation: in the oversampling unit M-1 samples out of *M* are zeros (due to zero padding) and hence a polyphase structure reduces the required MAC operations for the interpolating filter by a factor *M*.

2 Multistage cascade realization: the oversampling unit can be realized through a cascade of *S* multiple stages where the *i* th stage realizes an oversampling by a factor *Mi* (with $M1 \cdots Mi \cdots MS = M$) with a filter of reduced order $Z_i \ll Z$.

Table 4 shows the overall MAC computational cost, considering polyphase filter implementations, of different possible partitioning of the whole oversampling unit in multiple cascaded stages. The analysis in Table 4 considers an oversampling factor $M = 8$or 16 and $F_{IN} = 44.1$ kS/s. In case of DVD-audio with $F_{IN} = 96$kS/s, MAC computations and PWM frequency in Tables 3 and 4 are doubled.

Table 4. Interpolating filter complexity, polyphase, and multistage units.

Mask type	Filter type	Polyphase cost $10^6 \cdot$ MAC/s	Filter order, ith stage			
			$i = 1$	$i = 2$	$i = 3$	$i = 4$
		$M = 8$, $S = 3$ stages $x2 - x2 - x2$				
1	Equiripple	3.26	48	7	3	—
2	Equiripple	3.26	32	11	5	—
1	Kaiser	5.07	59	12	8	—
2	Kaiser	4.72	37	15	10	—
		$M = 16$, $S = 4$ stages $x2 - x2 - x2 - x2$				
1	Equiripple	4.32	48	7	3	3
2	Equiripple	4.32	32	11	5	3
1	Kaiser	7.54	59	12	8	7
2	Kaiser	7.89	37	15	10	9
		$M = 16$, $S = 3$ stages $x2 - x2 - x4$				
1	Equiripple	4.67	48	7	11	—
2	Equiripple	5.20	32	11	16	—
1	Kaiser	6.48	59	12	16	—
2	Kaiser	6.48	37	15	20	—

Comparing the results of Tables 3 and 4 it is clear that, exploiting the multistage decomposition and the poly-phase techniques, the filter complexity is reduced to few millions of MAC/s.

The computational burden in Table 4 is roughly the same for the two mask types; in the rest of the work mask 2 is used since it ensures a lower pass-band ripple. Using the Equiripple filter type a computational saving of roughly 30% is achieved versus the Kaiser type.

By adopting an oversampling factor $M = 16$, 4-stage polyphase FIR filter of Equiripple type with filter mask 2, the required complexity can be sustained for both CD-quality and DVD-quality inputs realizing in hardware a single MAC unit with 1 MAC/cycle capability and with a clock frequency below 10 MHz. The use of higher oversampling factors is still limited by the switching frequency of the power output bridge (see Section 3.6).

Finally, the bit-true arithmetic of the MAC hardware unit has been determined: using a 12-bit fixed-point data width the circuit complexity is greatly reduced versus a floating—point arithmetic implementation while the reduction of audio reproduction quality is negligible.

3.2. Noise Shaper

Converting the n-bit oversampled PCM signal to PWM leads to a minimum impulse time $T_{min} = 1 / (F_C \cdot 2^n)$ s, for example, roughly 0.25 nanoseconds considering $M = 16$and the 44.1 kS/s 16-bit PCM signal of audio CDs. Such values are too low for commercial power transistors [26, 27] with rise and fall times, T_r and T_f, of tens of nanoseconds. To reduce such requirement while keeping unaltered source audio quality, a noise shaper is used. Its architecture is shown in Figure 3: it reduces the used bits from n to p, while the added quantization noise can be spread outside the audio band using a Kthorder FIR shaping filter. In our model the noise shaper is parametric in terms of output bit width p and filter order K. An high value for K leads to improved SNR performances but also to an increased circuit complexity and to the risk of loop instability. Our analysis proves that stable loops and good trade-offs between audio performance and circuit complexity can be achieved using p between 4 and 8 bits and noise shapers with an FIR filter up to 5thorder. Figure 4 reports the magnitude response of the selected noise shaping filter for the prototyping phase described in Section 4. The noise shaper transfer function NTF is $\mathrm{NTF} = ens(z) / erq(z) = H(z) - 1$, being $H(z) = 1 - (1 - 1/z)^5$ the shaping filter response.

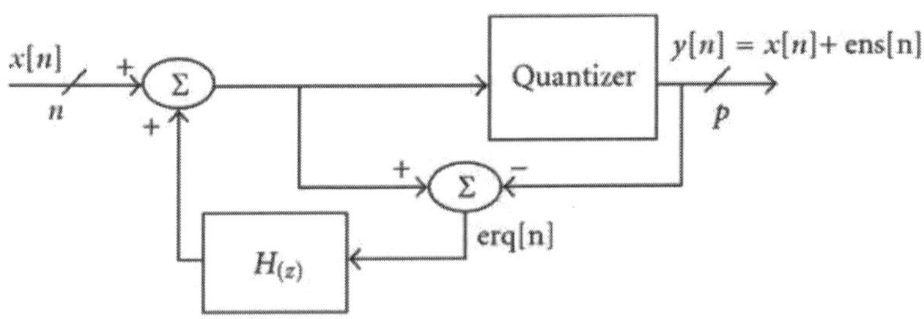

Figure 3. Noise shaping circuit (n bits input, p bits output).

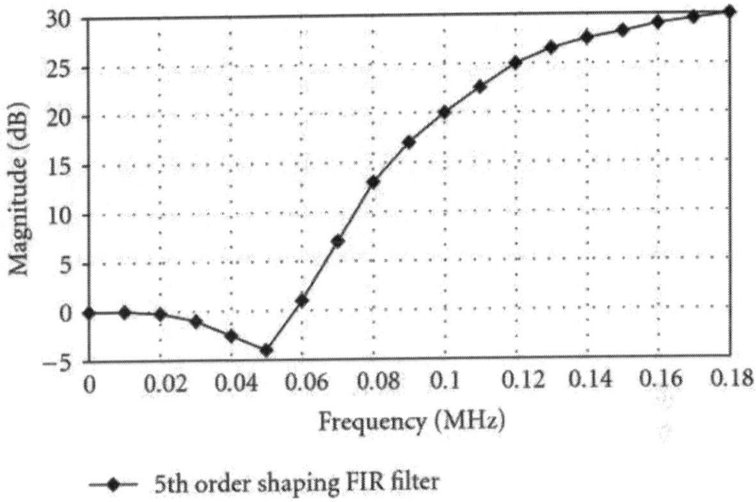

Figure 4. Noise shaping filter frequency response.

The most suited values for M and p depend on the time response of the used power MOS; the minimum impulse time $T_{min} = 1/(M \cdot F_{IN} \cdot 2^{p})$ should be comparable to the sum of T_r and T_f. As example for CD-audio signals, using $M = 16$, $p = 7$ leads to a T_{min} of 11 nanoseconds compatible with the timing of the selected power MOS devices [26].

3.3. Cross Point Estimator for NPWM

The scheme with oversampling plus noise shaping implements the Uniform sampled PWM (UPWM). To further reduce the THD a Natural PWM (NPWM) modulation can be realized. For NPWM, see Figures 2 and 5, a cross-point deriver is added after the oversampling unit to estimate the intersection point between the sawtooth carrier and the original modulating signal as in analog PWM. In our mixed-signal model we have implemented four cross point estimators with different performance-complexity trade-offs. Two high-performance algorithms to estimate the NPWM cross point have been derived from [22], based on the iterative application of a Newton-Rapson method, and from [13], based on a 4-point Lagrange interpolator. Our simulations prove that these cross point estimators can reduce the THD down to 0.02% but at the expense of high computational cost. Hundreds of millions of MAC operations per second are required for CD inputs and hence an extra DSP processor has to be dedicated to the cross-point estimation task.

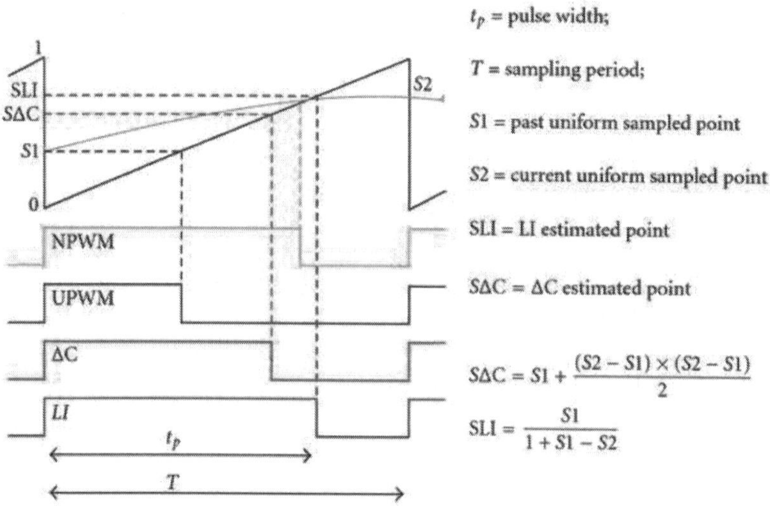

Figure 5. Example pulse width in UPWM, NPWM, ΔC- and LI- NPWM.

Simplified 2-point estimators for NPWM, using linear interpolation (LI) or delta compensation (ΔC), allow a single chip implementation of the whole signal processing part. Here the crossing-point is estimated using the first-order formulas reported in Figure 5 requiring just 1 multiply/sample for ΔCand 1 division/sample for LI. Differently from [14, 15], where LI is preferred to ΔCin case of micropower speech amplifiers not using oversampling, our simulations prove that the THD reduction using NPWM-LI or - ΔCis the same: a factor of 2 lower than UPWM. Thus, the use of ΔC is preferred since it is less complex than LI requiring the computation of a multiplication/sample instead of a division/sample.

3.4. Multilevel PWM

A classic PWM has $L = 2$levels but also $L = 3$levels PWM and $L > 3$PWM techniques, up to 9 in [16], can be implemented. In 2-state PWM the signal is switching between maximum and minimum supply voltage values, the two states $V^+$and V^-. Even for low-level signals binary modulation continuously provides energy to the LC filter and to the load. If the modulating input is null the PWM wave is still switching with a 50% duty cycle. The signal provided to the load is null but switching losses are paid thus reducing power efficiency. The 3-state PWM signal switches between $V^+$and 0 when the input signal is positive, otherwise between $V^-$and 0. In case of null input there is no switching activity and hence switching losses are reduced. The 3-state PWM modulator reduces by a factor 2 the voltage swing supported by the power MOS transistors allowing also the reduction of electromagnetic interference (EMI) and a better behaviour of the power devices. By further increasing the number of levels *L* the EMI and the switching losses can be further reduced. However while 2 and 3 levels PWM can be implemented with a single power MOS bridge, a PWM with levels $L > 3$ require multiple power bridges, with matched behaviour. This increases remarkably the amplifier complexity and cost. Our analysis proves that for $L > 3$the complexity increase is not justified by a performance gain which is limited if compared to ternary PWM.

While the power output stage can be the same for 2 and 3 levels PWM, the digital circuitry is different. To generate a 2-state PWM digital wave each *p*-bit noise shaped and oversampled sample is compared to a digital sawtooth waveform. The 3-state PWM modulator is realized using two 2-state PWM modulators, one for positive input samples and the other for negative samples: after controlling the sign of each sample only one of the two 2-state modulators is enabled. Although 3-state PWM modulation doubles the cost of the 2-state one the complexity of the whole system is comparable. Indeed, as shown in Section 4, the overall digital circuitry complexity is dominated by the noise shaping and oversampling filters which are common to 3- and 2-state PWM. Summarizing 3 levels PWM is the best trade-off between performance and circuit complexity.

3.5. Dead Time Insertion

Before driving the power stage, proper guard intervals have to be inserted at the beginning and at the end of each PWM word to take into account the switching delay time of the selected power MOS devices. The minimal time resolution of the PWM wave is $T_{min} = 1 / (M \cdot F_{IN} \cdot 2^P)$, determined by the oversampling and noise shaping choices, which can be smaller than the switching transition time of typical power MOS. As example the sizing in Section 3.2 leads to a T_{min} of 11 nanoseconds smaller than the 20 nanoseconds switching delay of the selected MOS devices [26]. If the time guard intervals are not inserted, PWM words with duty cycles of few % or near 100%, that is, with high or low time intervals of tens of nanoseconds, cannot be correctly managed by the power stage and distortions will arise. Moreover, since the switching times of P- and N- MOS are not the same, for each transition of the PWM signal there is the risk of a short circuit between the voltage supplies (a MOS is already on while the other is not completely off). To avoid these power wasting phenomena extra dead time intervals have to be inserted when the PWM wave is switching. On the other hand the higher the inserted time guard and dead time intervals, the higher the reduction of the amplifier dynamic range. An optimal sizing of the time intervals to be inserted can be found simulating in the Simulink/Spice environment the whole amplifier including accurate models of the power MOS time response. Figure 6 shows the THD reduction factor, that is, the ratio between the THD obtained when inserting different guard intervals *tg* versus the THD obtained without any time guard. In Figure 6 the THD is reduced up to a factor 4 using a *tg* of 30 nanoseconds (curve "typical"). The example architecture considered in Figure 6 is that with oversampling and noise shaping sized as in Section 3.2 plus ternary PWM, power stage of Section 3.6, no feedback loop. The optimal value for *tg* in Figure 6 is derived from architecture simulations; a mathematical formula in closed form is not available. When considering PVT (process, voltage and temperature) variations of the implementing hardware devices (e.g., the FPGA and the power components of the prototype in Section 4) the THD reduction curve varies; see as example in Figure 6 the curves reported with dashed lines corresponding to the "min" and "max" corner cases. Using the *tg* value calculated in nominal conditions (30 nanoseconds in our case, see "typ" curve in Figure 6) a suboptimal, but still noticeable, THD improving factor is obtained versus the case with no time guard insertion. In Figure 6 with *tg* = 30 nanoseconds the THD reduction versus the case without any time guard is from 0.25 to 0.55 depending on the PVT conditions. Moreover, as discussed in Section 3.5, the use of a proper feedback scheme can reduce the architecture sensitivity to parameters variation.

3.6. Output Power Stage

Figure 7 reports the circuit schematic of the full bridge power stage, using commercial power MOS from [26]. A complementary N-/P-MOS solution is adopted. A level-shifting circuitry is used for high-side gate driving. Discrete power MOS are available with associated gate driver buffers (represented by B1, B2, B3, B4 in Figure 7), allowing the connection of the output power stage to the PWM output of a low-power digital circuit, such as an FPGA. For the target power levels of this work the supply voltage *Vdd* in Figure 7 is sized at 25 V. The LC filter in Figure 7 is a differential 4th-order Butterworth with 20 kHz cutoff. In our model we compared several LC filtering stages, with different filter order and considering Butterworth, Chebychev and Elliptic types.

Figure 8 reports the THD achieved using Butterworth, Chebychev (0.5 dB pass-band ripple) and Elliptic (0.1 dB pass-band ripple) analog filters with a cutoff frequency of 20 kHz and considering different filter orders. From Figure 8 it can be noted that the performances of the filters are similar (elliptic filters are not defined for orders lower than 3). Butterworth type is preferred to avoid ripple in the pass-band. After the 4th order the THD reduction obtained increasing the filter complexity is minor.

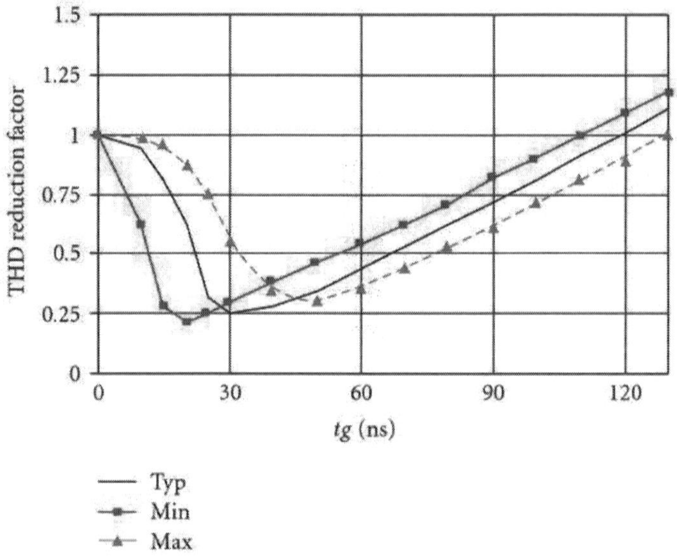

Figure 6 .THD reduction versus inserted time guard tg.

3.7. Feedback Topologies

As far as feedback topology is concerned different schemes have been modelled. One solution is an open loop amplifier avoiding the problem of how to reinsert the power output signal in the low-power digital processing chain. This solution is widely adopted in literature, for example, in [13, 19]. Our simulations prove that an open loop scheme adopting oversampling, noise shaping, ΔC cross point estimation, 3-level PWM and dead time compensation (this scheme is nicknamed A1 in Figures 9 and 10) can be properly configured taking into account the real characteristics of the power stage. This open-loop amplifier leads to optimal THD and power efficiency results, see Figures 9 and 10, comparable to those obtained with closed-loop feedback amplifiers but avoiding extra hardware components. However if a circuit parameter changes (e.g., the PVT variations discussed in Section 3.4 and Figure 6) versus the reference value used for configuration or the supply voltage is affected by ripple there is not a compensating mechanism. Indeed, for open loop amplifiers it is mandatory the use of regulated supply voltages [19]. To reduce the amplifier sensitivity a feedback loop can be added to the architecture configuration A1. As example, in the architecture nicknamed A2 in Figures 9 and 10, the signal generated by the digital PWM modulator is compared to a scaled version of the output amplified PWM wave. Their difference *Verr* is sent to a lowpass analog controller, with transfer function C(s), extracting the DC component of the error. In our model we considered and simulated for C(s) an active filter (OpAmp plus RC network) with parametric cutoff frequency and filter order. The generated DC error level is then used in the PWM correction unit of Figure 2 to properly insert time delays in the PWM modulator thus driving the power MOS stage with a corrected PWM signal. Note that such approach is similar to some feedback algorithms proposed in literature: for example, in [9] C(s) is realized as a simple 1storder integrator while in [11] a more performing algorithm called PEDEC (Pulse Edge Delay Error Control) is used for PWM correction, starting from the generated DC error level. The results in Figures 9 and 10 for the amplifier nicknamed A2 refer to a feedback configuration with a 3rdorder lowpass C(s) filter and using PEDEC as compensating technique. This approach provides good results but is not useful for a low-complexity and low-cost realization since it requires an extra analog feedback network.

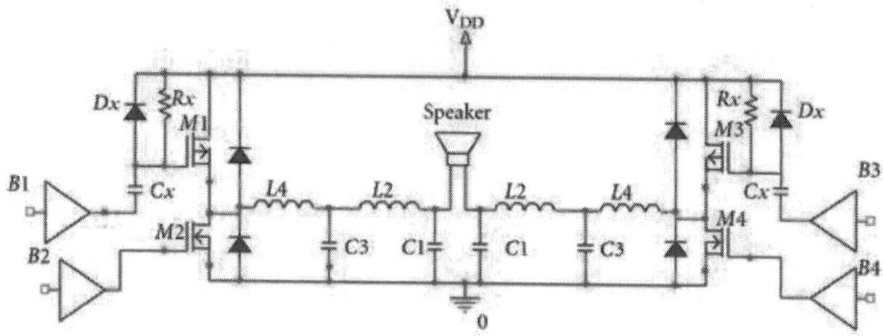

Figure 7. Full bridge power stage schematic.

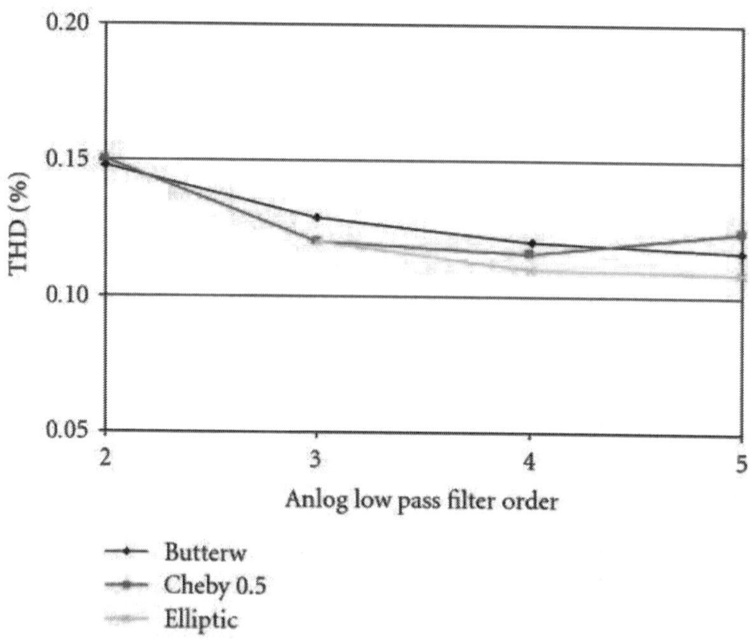

Figure 8. THD reduction for different analog LC filter types.

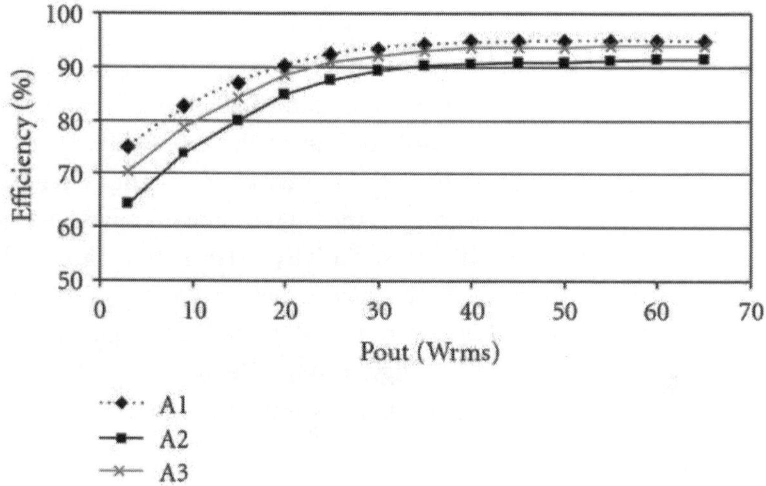

Figure 9. Feedback type comparison, power efficiency versus Pout.

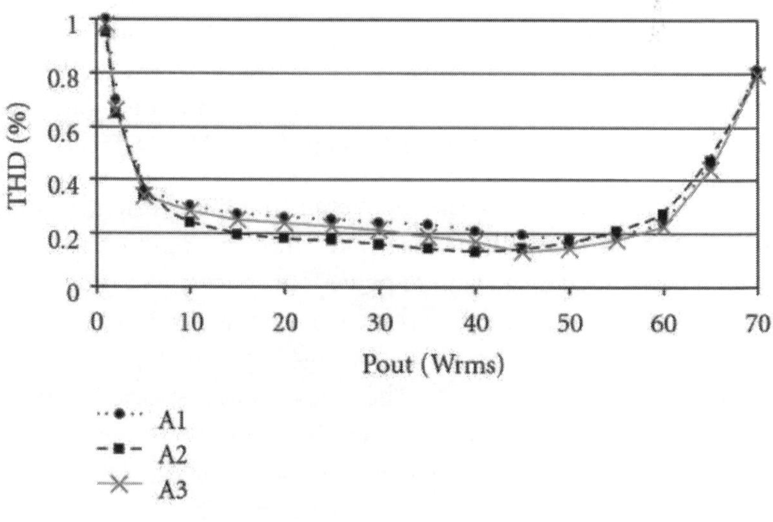

Figure 10. Feedback type comparison, % THD (at 1 kHz) versus Pout.

With respect to that adopted in A2, more complex mixed-signal feedback correction schemes have been proposed in literature [10, 18]. In these schemes the output PWM power signal, after attenuation and filtering, is converted in the digital domain through extra ADCs: 7-bit flash ADC in [10] and 10-bit SAR ADC in [18]. In the digital domain similar operations to those of A2 are carried out. These feedback schemes using ADC and operating in the digital domain have been also modelled. The achievable THD and sensitivity performances are slightly better than those of A2 while the hardware overhead is much higher: an attenuator, a filter and an ADC are required in the analog domain plus a digital correcting unit in the digital one. Being too complex the feedbacks schemes with ADCs are not considered in the comparison of Figures 9 and 10.

A simpler but effective feedback correction technique to reduce the amplifier sensitivity to parameter changes is derived from [21]: the sign of the output current provided to the load is used as 1-bit control to check periodically which output transistor is on and to change consequently the inserted time-guard value. In [21] this technique has been proposed for the control of a 2-state PWM power-bridge. In this work this approach has been redesigned to be integrated with oversampling, noise shaping, ΔC cross point estimation and 3-level PWM creating a new amplifier scheme: A3, 1-bit feedback extension of the open loop A1. In our implementation we selected the value of 10 nanoseconds as resolution of correction for the PWM waveform.

With reference to max 70 Wrms delivered to a 4 Ωspeaker, Figures 9 and 10 compare the amplifier schemes A1, A2 and A3 in terms of power efficiency and THD. The maximum power efficiency, up to roughly 95% in Figure 9, is achieved by the open loop scheme A1. With such high-efficiency each power MOS of the full bridge is dissipating less than 1 W avoiding extra cooling hardware. These results outperform classic DAC plus analog amplifier solutions: as example the hybrid analog scheme in [5] has a maximum power efficiency below 77%. The schemes with feedback topologies, A2 and A3, achieve similar efficiency, higher than 90%, only for power levels higher than 25–30 Wrms. Concerning THD, the lower distortion is achieved around 40–50 Wrms. The minimum THD is below the target of 0.2%; the use of feedback schemes allows reaching the target THD on a wider frequency range versus the open loop A1 scheme. Between A2 and A3 amplifiers the latter is preferred since it improves the THD and sensitivity performance of A1 but with minimal complexity overhead and minimal efficiency losses.

4. PROTOTYPED DIGITAL AUDIO AMPLIFIER

From the design space exploration carried out in Section 3 the amplifier architecture, summarized hereafter, resulted as an optimal trade-off between circuit complexity, power efficiency, output distortion and low sensitivity to parameter changes. The digital part includes: an oversampler by a factor $M = 16$ using an Equiripple 4-stage polyphase FIR interpolating filter, a cross point estimator based on ΔC technique realizing a NPWM scheme, a noise shaper with $P = 7$ output bits and a 5th-order noise shaping filter, 3-level PWM generation, correction of PWM words through the insertion of time guard intervals also as a function of a 1-bit signal feedback.

The digital processing part, implemented in HDL, has been synthesized in different CMOS standard-cells technologies (90 nm 1 V supply voltage and 180 nm 1.8 V supply voltage) resulting in a digital complexity of 15.2 Kgates, mainly due to the noise shaping and interpolating filters. The low circuit complexity allows the fitting of the digital circuitry on several low-cost SRAM-based FPGA devices. As example the processing part of the amplifier occupies 90% of a Xilinx Virtex XCV100 or 58% of a Xilinx Spartan3 200. Such devices are available for large volume production at a cost of few dollars. Hence the low circuit complexity of the proposed architecture allows for a low-cost implementation. The power consumption for the above cited implementations is in the order of few hundreds of mW, as example 100 mW when integrating the amplifier in the XCV100 and playing 44.1 kS/s CD-quality audio signals.

The output stage is a full bridge made up of N-/P- MOS devices from [26] plus a differential 4th-order LC Butterworth filter. Table 5 summarizes the measured results on the amplifier prototype. The THD value at 1 kHz refers to a 16-bit 44.1 kS/s CD-quality input signal and is evaluated using an Audio Precision test setup. The high power efficiency achieved permits an output power up to 70 Wrms without using extra cooling hardware. The results in Table 5 (Pmax of 70 Wrms, THD of 0.13% and efficiency of 94% at 45 Wrms) confirm the performance estimation made by simulations during the design phase in Sections 2 and 3, particularly in Figures 9 and 10 (Pmax of 70 Wrms, THD below 2% and efficiency up to 95% in the range 40–50 Wrms).

Table 5. Summary of amplifier characteristics.

Design metric	Value	Test conditions
Efficiency	94% at 45 Wrms	16-bit 44.1 kS/s signal, 1 kHz tone
THD	0.13% at 45 Wrms	
Pmax	70 Wrms	4 Ω load

The prototype allowed us also to assess the performance improvement of the 1-bit compensating loop scheme. The sensitivity of the amplifier to parameter changes is an important feature against temperature variations, devices tolerances, power supply ripple. As example, Figure 11 shows how the THD degrades when the system configuration is optimised for 4 Ωspeaker and then the load is changed with an 8 Ω speaker. Both the cases with 1-bit feedback scheme (*1b feedback*) and without (*open loop*) are considered. From Figure 11 it is worth noting how the THD performance degradation due to parameter change is lower in the amplifier with feedback.

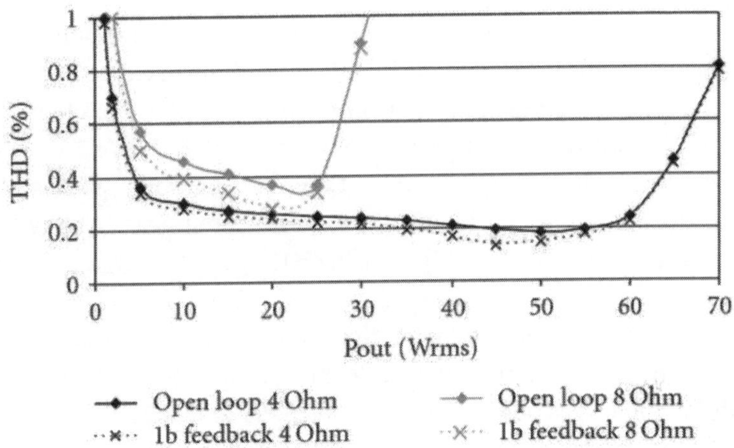

Figure 11. Architecture sensitivity example, % THD in case of load change.

5. COMPARISON TO THE STATE OF THE ART AND FUTURE WORK

5.1. Comparison to the State of the Art

When compared to the state of the art of digital input power amplifiers our prototype stands for its low-complexity, while keeping high power efficiency and low THD levels. The distortion levels, THD below 0.2% at 1 kHz in the power range 35–60 Wrms with a minimum of 0.13% at 45 Wrms (see A3 in Figure 10), are suitable for audio Hi-Fi applications. Other works in literature achieve lower THD values, as example in [13] the THD is 0.02% for similar power levels of 50 W, but at the expense of a lower efficiency and an increased complexity. The power efficiency is 80% in [13] while in our work is higher than 90%. The digital processing tasks in [13] require the use of multiple boards (1 DSP board for the digital audio processing plus 1 FPGA board for PWM processing) while our architecture just requires 1 low-cost FPGA having a bounded circuit complexity of 15 Kgates. The multiple boards digital amplifier in [13] features also a configuration with 90% power efficiency but with a THD of 0.2%. In [16] with a 9 level PWM inverter the achieved performances of 0.25% THD and 80% efficiency are worse than our results. This confirms our analysis in Section 3.4 that 3-level PWM is the optimal choice for the output stage. Compact solutions using a single chip for the digital part, as in our work, and without heatsink have been proposed in [9, 19, 23, 31, 32]. However [19] is missing the feedback scheme needed, as proved in Section 4, against parameter changes; in [9] the efficiency levels are lower than those achieved by our scheme. The FPGA-based audio amplifier in [23] is missing feedback and NPWM techniques; it has a power efficiency of 80% and a THD of 1% both worse than our achieved results. In [31, 32] only the interpolation filter is implemented occupying a whole Spartan FPGA. With respect to our previous conference paper [20], where only UPWM is implemented and a lower oversampling factor and a less-performance FIR interpolating filter are used, the audio processing system in this work has been improved including the digital techniques for NPWM, more accurate models for all the analog components, a more performing interpolating filter. The prototyped architecture in this work versus [20] achieves a much better THD value, predicted by the simulations and confirmed by measures on the prototype. Finally in some works [10, 17, 24], the shown results refer to simulations or prototypes only of the low-power PWM generator without including a real prototyped power stage. As discussed in this work, and widely proved in literature, the nonideal behaviour of the power stage is a key issue in power audio amplifiers.

5.2. Future Work

The proposed platform-based approach has been used also to define the optimal architecture of digital power audio amplifiers using other complementary power MOS devices, such as the IR530 and IR9530. The achieved results with these MOS devices prove that, targeting a power level of 45 Wrms on an 8 Ωload, optimal distortion performances below

0.2% can be reached in the range 17–35 Wrms with a power efficiency higher than 90%. The resulting architecture is similar to that discussed in Section 4 with the exception of the tuning of some parameters specifically optimized for the characteristics of the new power devices.

As work extension we are applying the same methodology to the design of a fully integrated digital input audio amplifier targeting maximum power levels of 1-2 W. Such amplifiers of few Watts are needed for battery-power terminals with audio playing capability [33, 34]. The design of the amplifier is carried out using an architecture similar to that in Section 4 fitted on a BCD 0.35 μm technology providing CMOS transistors for the digital part and DMOS transistors for the analog power part. The only off-chip circuit is the LC lowpass filter. Postlayout characterization proves that the digital amplifier can be integrated in less than 2 mm^2. The integrated power stage is an inverter with NDMOS sized with W = 22mm and L = 6μm supporting, with low R_{DSon} of few mΩ, output currents of 0.14 A on output load of 100 Ω.

6. CONCLUSIONS

The design of digital audio power amplifiers is presented in the paper. A modelling platform has been built to allow a fast but still accurate exploration of the mixed-signal design space which involves the codesign of (i) audio processing algorithms with physical characteristics of hardware components and of (ii) low-power integrated digital circuits with analog power devices. Different amplifier architectures have been modelled, simulated and compared to find optimal trade-offs among different cost-functions: low-distortion, high power efficiency, low circuit complexity and low sensitivity to parameter changes. The selected amplifier architecture has been prototyped, for power levels of tens of Watts, implementing the digital processing part on a single low-complexity FPGA while off-chip components are used for the power output stage, no heatsink is required. The resulting digital amplifier, compared with the state of the art, features a low circuit complexity while keeping good power efficiency, higher than 90%, and low-distortion levels, down to 0.13%. As future extension the realization of a fully integrated digital amplifier in BCD technology is presented for power levels of few Watts.

ACKNOWLEDGMENT

The work has been partially supported by the SHAPES FP6 EU project.

REFERENCES

1. Self D: Audio Power Amplifier Design Handbook. 4th edition. Newnes, Oxford, UK; 2006.

2. Saponara S: **Current-feedback architecture for high-slew-rate and low-THD high-end audio amplifier.** Electronics Letters 2008,**44**(25):1433-1434. 10.1049/el:20082660

3. Walker G: **A class B switch-mode assisted linear amplifier.** IEEE Transactions on Power Electronics 2003,**18**(6):1278-1285. 10.1109/TPEL.2003.818825

4. Ginart AE, Bass RM, Leach WM Jr., Habetler TG: **Analysis of the class AD audio amplifier including hysteresis effects.** IEEE Transactions on Power Electronics 2003,**18**(2):679-685. 10.1109/TPEL.2003.809330

5. Saponara S, Terreni P: **Switching-based topologies for high-efficiency audio amplifiers.** Proceedings of the International Symposium on Signals, Circuits and Systems (ISSCS '05), 2005 **1:** 283-286.

6. Poulsen S, Andersen M: **Hysteresis controller with constant switching frequency.** IEEE Transactions on Consumer Electronics 2005,**51**(2):688-693. 10.1109/TCE.2005.1468020

7. Hiorns RE, Sandler MB: **Power digital to analogue conversion using pulse width modulation and digital signal processing.** IEE Proceedings G 1993,**140**(5):329-338.

8. Botti E, Grosso A, Meroni C, Stefani F: **Digital input audio power amplifiers in 0.6- μ m BCD technology: two examples.** Proceedings of the IEEE International Symposium on Power Semiconductor Devices & ICs (ISPSD '04), 2004, Kitakyushu, Japan **16:** 93-96.

9. Grosso A, Botti E, Stefani F, Ghioni M: **A 250 W audio amplifier with straightforward digital input-PWM output.** Proceedings of the European Solid-State Circuits Conference (ESSCIRC '05), 2002 225-228.

10. Midya P, Roeckner B, Bergstedt S: **Digital correction of PWM switching amplifiers.** IEEE Power Electronics Letters 2004,**2**(2):68-72. 10.1109/LPEL.2004.834795

11. Nielsen K: **PEDEC—a novel pulse referenced control method for high quality digital PWM switching power amplification.** Proceedings of IEEE Power Electronics Specialist Conference (PESC '95), 1998 200-208.

12. Saponara S, Fanucci L, Terreni P: **Oversampled and noise-shaped pulse-width modulator for high-fidelity digital audio amplifier.** Proceedings of the IEEE International Conference on Electronics, Circuits, and Systems (ICECS '06), 2006, Nice, France 830-833.

13. Pascual C, Song Z, Krein PT, Sarwate DV, Midya P, Roeckner WBJ: **High-fidelity PWM inverter for digital audio amplification: spectral analysis, real-time DSP implementation, and results.** IEEE Transactions on Power Electronics 2003,**18**(1):473-485. 10.1109/TPEL.2002.807102

14. Gwee B-H, Chang JS, Adrian V: **A micropower low-distortion digital class-D amplifier based on an algorithmic pulse width modulator.** IEEE Transactions on Circuits and Systems I 2005,**52**(9):934-949.

15. Gwee B: **Micropower low-distortion digital pulse width modulator for a digital class D amplifier.** IEEE Transactions on Circuits and Systems II 2002,**49**(5):1-13.

16. Antunes VME, Silva JF, Fernão-Pires V: **Experimental evaluation of a digital multi-level audio power amplifier.** Proceedings of the IEEE Annual Power Electronics Specialists Conference (PESC '04), 2004 **2:** 1175-1179.

17. Vlassopoulos N, Reisis D, Lentaris G, et al.: **An approach for efficient design of digital amplifiers.** Proceedings of the IEEE International Symposium on Circuits and Systems (ISCAS '06), 2006 5531-5534.

18. Spectrum Design Solutions Inc : **Digital audio solutions application brief.** 2006.

19. Apogee : **Direct digital amplification (DDX) white paper.** 2002.

20. Saponara S, Terreni P: **Mixed-signal design of a digital input power amplifier for automotive audio applications.** Proceedings of the Conference on Design, Automation and Test in Europe (DATE '06), 2006 **2:** 212-216.

21. Jeong S-G, Park M-H: **The analysis and compensation of dead time effects in PWM inverters.** IEEE Transactions on Industrial Electronics 1991,**38**(2):108-114. 10.1109/41.88903

22. Goldberg JM, Sandler MB: **Pseudo-natural pulse width modulation for high accuracy digital-to-analogue conversion.** Electronics Letters 1991,**27**(16):1491-1492. 10.1049/el:19910933

23. Liu Y-H, Teng J-H, Hsieh C-H: **Design and implementation of a fully-digital, high-efficiency class-D amplifier system.**

Proceedings of the IEEE Region 10th Annual International Conference (TENCON '07), 2007, Taipei, Taiwan 1-4.

24. Yoneya A: **Pulse width and position modulation for fully digital audio amplifier.** Proceedings of the IEEE International Symposium on Circuits and Systems (ISCAS '08), 2008, Seattle, Wash, USA 1692-1695.

25. Saponara S, Nuzzo P, Nani C, van der Plas G, Fanucci L: **Architectural exploration and design of time-interleaved SAR arrays for low-power and high speed A/D converters.** IEICE Transactions on Electronics 2009, **E-92C:** 843-851.

26. Infineon Technologies : **Data sheet BSO200N03S, Optimos 2 Power-Transistor.** rev 1.6, 2008

27. Cerezo J: **Class D audio amplifier performance relationship to MOSFET parameters.** International Rectifier AN 1070, 2005

28. Philips : **Data sheet UDA1320ATS.** January 2000

29. Turek D: Design of efficient digital interpolation filters for integer upsampling, M.S. thesis. MIT, Boston, Mass, USA; 2004.

30. Benameur N, Loulou M: **Design of efficient digital interpolation filters and sigma-delta modulator for audio DAC.** Proceedings of the IEEE 4th International Conference on Design & Technology of Integrated Systems in Nanoscal Era, (DTIS '09), 2008 1-7.

31. Jacobsohn P, Palmer D: **Audio sample rate conversion in FPGAs.** Xcell Journal 2007, 50-54.

32. Huang X, Han Y, Chen L: **The design and FPGA verification of a general structure, area-optimized interpolation filter used in sigma-delta DAC.** Proceedings of the IEEE International Conference on Solid-State and Integrated-Circuit Technology (ICSICT '06), October 2006, Shanghai, China 2111-2113.

33. Becker R, Groeneweg WH: **An audio amplifier providing up to 1 Watt in standard digital 90-nm CMOS.** IEEE Journal of Solid-State Circuits 2006,**41**(7):1648-1653. 10.1109/JSSC.2006.873674

34. Ryoo J-Y, Cho G-H: **A single chip l W CMOS audio power amplifier with pseudo buffer analog and class D switching mixed mode for mobile application.** IEICE Transactions on Electronics 2005,**E-88C**(9):1886-1892.

CHAPTER 4

Hardware-Efficient Delta Sigma-Based Digital Signal Processing Circuits for the Internet-of-Things

Yifei Liu, Paul M. Furth and Wei Tang *

Klipsch School of Electrical and Computer Engineering, New Mexico State University, 1125 Frenger Mall, Las Cruces, NM 88003, USA

ABSTRACT

This paper presents hardware-efficient Delta Sigma linear processing circuits for the next generation low-power VLSI devices in the Internet-of-things (IoT). We first propose the P-N (positive-negative) pair method to manipulate both the analog value and length of a first-order Delta Sigma bit sequence. We then present a binary counter method. Based on these methods, we develop Delta Sigma domain on-the-fly digital signal-processing circuits: the Delta Sigma sum adder, average adder and coefficient multiplier. The counter-based average adder can work with both first-order and higher-order Delta Sigma modulators and can also be used as a coefficient multiplier. The functionalities of the proposed circuits are verified by MATLAB simulation and FPGA implementation. We also compare the area and power between the proposed Delta Sigma adders and a conventional multi-bit adder by synthesizing both circuits in the IBM 0.18-μm technology. Synthesis results show that the proposed Delta Sigma processing circuits can extensively reduce circuit area and power. With 100 inputs, a Delta Sigma average adder saves 94% of the silicon area and 96% of the power compared to a multi-bit binary adder. The proposed circuits have the potential to be widely used in future IoT circuits.

KEYWORDS

VLSI; Delta Sigma modulation; digital signal processing; adder; coefficient multiplier; low-power low-complexity circuits

1. INTRODUCTION

The Internet-of-Things (IoT) demands ubiquitous devices that collect and process sensed data and transmit back to the central node through various network interfaces [1]. Since radio devices usually dominate the power consumption of the system [2], in biomedical or physiological monitoring, the sensor data at each node should be preprocessed before transmission [3]. In particular, biomedical sensors in the IoT targets an important application of the "unmanned medical nursing system" [1], which could alleviate increasing medical costs as the baby boomers enter retirement [4]. Such wearable and implantable biomedical sensors demand low-complexity, low-power, high-resolution and high-reliability digital signal processing (DSP) circuits, which present the need to enhance the currently-applied solution. In a typical biomedical sensor, the recorded analog signal is first converted into digital format and then processed by an application-specific integrated circuit (ASIC) processor running a pre-implemented DSP algorithm, as shown in Figure 1a. If the algorithm determines that the incoming signal is abnormal, then the ASIC sends an alert signal and calls for help through wireless portable devices via the IoT. In such a system, the major challenges are the relatively large circuit area and high power of the DSP circuits performing multiply-and-accumulate (MAC) operations [5], which are adders and coefficient multipliers in digital filters. In order to miniaturize the physical size and extend the battery life of the circuits, low-complexity low-power designs are very important. As an example, in [5], an eight-channel 10-bit resolution DSP seizure detector processor consumes more than 50% of the chip area and power in a 5 mm × 5 mm system-on-a-chip (SOC) implemented in a 65-nm process, even though a look-up table (LUT) method is applied, which saved more than 50% of the total gate count in the digital filters [6]. Besides circuit complexity and power consumption, such circuits demand higher resolution and reliability, since higher resolution improves accuracy, and reliability is the key issue in biomedical applications. However, if a higher resolution circuit is implemented, then the power, chip area and cost of the DSP circuits also increase.

Delta Sigma technology can be an attractive solution for the aforementioned challenges and has been widely applied in biomedical circuits and systems [7,8,9,10,11,12] because of its unique properties. First, it is well known that a Delta Sigma analog-to-digital converter (ADC) has much less circuit complexity and lower power consumption than multi-bit ADCs. Second, in the Delta Sigma domain, the instantaneous analog value is represented by the moving average of the Delta Sigma bit-stream. Therefore, a few bit errors in the bit-stream do not disturb the signal level greatly [13]. Therefore, Delta Sigma systems have a higher tolerance to bit errors than multi-bit binary systems, which means Delta Sigma systems have higher reliability. Third, an oversampling Delta Sigma ADC can achieve a higher resolution (up to 24 bits) compared to a Nyquist rate multi-bit ADC (usually less than 18-bits). The Delta Sigma ADCs also relax the requirements on the analog anti-aliasing filter. Moreover, the resolution of a Delta Sigma system can be flexibly changed by adjusting its clock rate, without modifying the hardware. On the other hand, concerns for using a Delta Sigma system usually arise from its higher sampling rate and larger data size. However, in biomedical applications, the input signal usually has a bandwidth less than 2 kHz. Furthermore, the storage of raw data is usually neither necessary nor easily implementable. Thus, in such applications, using Delta Sigma bit-streams to represent and process information is both applicable and desirable.

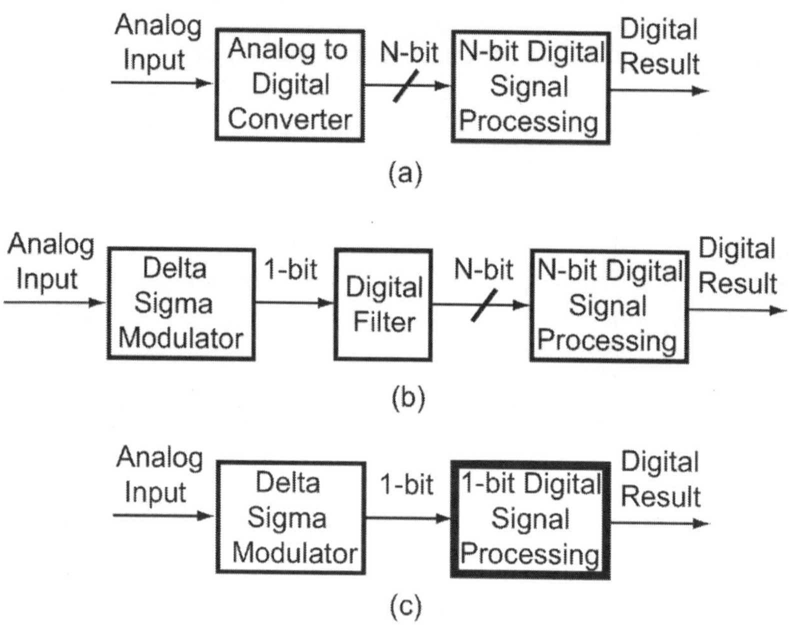

Figure 1. (a) Traditional signal processing architecture; (b) traditional signal processing after a Delta Sigma ADC; (c) the proposed signal processing using a Delta Sigma digital signal processing (DSP).

Although Delta Sigma technology has been widely used in converting analog signals to digital signals [7,9,11], it has not been commonly applied in DSP, because of the lack of Delta Sigma-based DSP circuits. Currently, the modulated Delta Sigma bit-stream is demodulated into a multi-bit stream and then processed by conventional multi-bit DSP circuits, as shown in Figure 1b [14]. However, hardware complexity, power and reliability issues limit the performance of the multi-bit DSP system. Here, we study the alternative method using a one-bit DSP system to process Delta Sigma bit-streams directly, as shown in Figure 1c.

Currently, Delta Sigma-based DSP circuits are not sufficiently studied. In DSP systems, the most important linear circuits are sum adders, average adders and coefficient multipliers. To build those circuits with low noise and complexity, pure digital implementations are preferable. However, some previous Delta Sigma processing methods focused on modifying the Delta Sigma modulators [15,16,17] instead of processing the modulated bit-stream. Moreover, some solutions are hindered by involving analog operations [18,19,20] or tri-state digital circuits [21], which increase the complexity of the circuits. Other designs are impeded by their ability to process only first-order Delta Sigma bit-streams [22,23,24,25,26,27,28,29,30,31,32,33], which do not fit the widely-used higher-order Delta Sigma modulators. Furthermore, the adders in [22] can only take two inputs. One method of processing higher order Delta Sigma bit-streams is the digital Delta Sigma modulator proposed. However, a digital Delta Sigma modulator designed with a complicated sorting network [13] requires enormous circuit resources, which is not applicable for low-power devices.

In this paper, we propose novel methods and circuits for Delta Sigma-based DSP systems. Based on the digital Delta Sigma modulator method and the proposed P-N (positive-negative) pair method with a signed binary counter, we design the Delta Sigma sum adder, average adder and coefficient multiplier. We also present simulation and measurement results of the proposed circuits and provide a comparison of the synthesis results between Delta Sigma adders and multi-bit binary adders. This paper is organized as follows: Section 2 presents the properties of a first-order Delta Sigma modulated bit-stream and the basic processing techniques using the proposed P-N pair method. In Section 3, we describe the design of the Delta Sigma sum adder and the average adder. Section 4 provides simulation and analysis of the Delta Sigma adders. Section 5 describes a Delta Sigma compressor design. Section 6 presents the implementation and measurement results of the Delta Sigma average adder. Section 7 compares the proposed Delta Sigma adders to conventional multi-bit adders. Section 8 concludes the paper.

2. DELTA SIGMA BIT-STREAM AND BASIC SIGNAL PROCESSING TECHNIQUES

In this section, we first briefly explain the properties of a first-order Delta Sigma-modulated bit-stream. Based on these properties, we propose the P-N pair method to process the Delta Sigma-modulated bit-streams.

2.1. Properties of First-Order Delta Sigma-Modulated Bit-Streams

A Delta Sigma modulator converts an analog input signal into a Delta Sigma-modulated digital bit-stream. A first-order Delta Sigma modulator consists of an analog integrator, a one-bit ADC, and a one-bit digital-to-analog converter (DAC), as shown in Figure 2. In the Delta Sigma modulator circuit, the analog input port and the digital output port can use either dual power supplies or a single power supply. To avoid confusion, we use P and N to represent the polarity of the output digital signals. P means a digital positive value of "true", while N means a digital negative value of "false". We use numbers to represent digital or analog levels. For example, −1 and +1 represent the dual power supply level VSS and VDD, respectively. Zero means analog ground, which is zero volts in a dual power system.

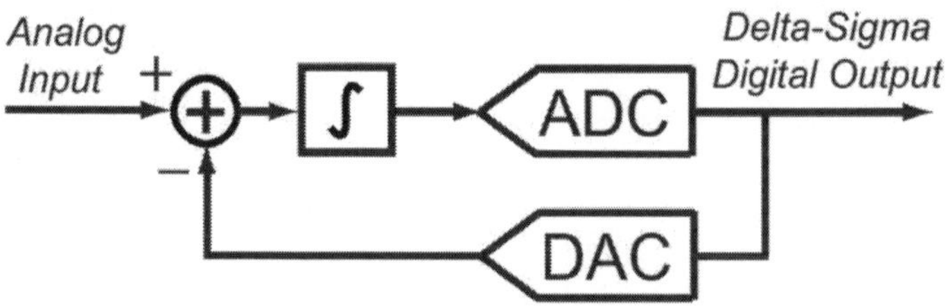

Figure 2. Schematic diagram of a first-order Delta Sigma modulator.

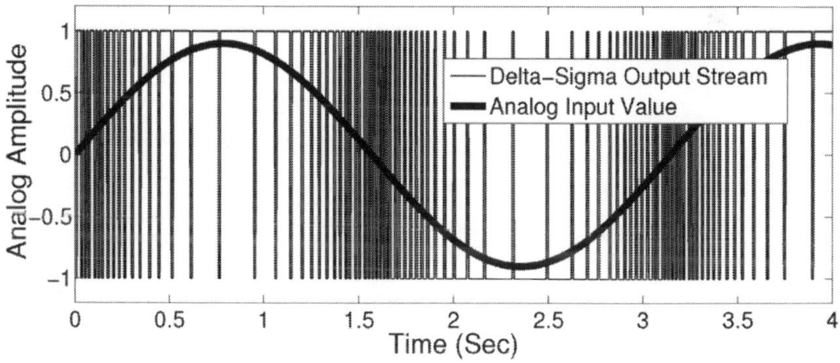

Figure 3. Waveforms of the input analog signal and the output Delta Sigma-modulated bit-stream of a first-order Delta Sigma modulator.

Figure 3 shows the input analog waveform and the output digital waveform of a first-order Delta Sigma modulator. When the input is positive, the digital output bit-stream contains more P bits than N bits. There can be consecutive P bits, but each N bit is separated by at least one P bit. Thus, there are no adjacent N bits when the input is positive. Similarly, when the input is negative, the output bit-stream consists of single P bits separated by one or more N bits. A Delta Sigma bit-stream toggles equally between N bits and P bits when its analog value is zero. Delta Sigma modulation is a type of pulse density modulation (PDM), which means the output digital bit-stream can be converted back to an analog value using a digital moving-average filter. Therefore, suppose

Delta Sigma-modulated bit sequence a has a total of W bits, if n of which have the value N and p of which have the value P. Then the analog value y of bit sequence a can be written as:

$$y = \frac{1}{W}\sum_{i=0}^{W-1} a_i = \frac{-n+p}{W} \tag{1}$$

where W = n + p. For any bit sequence of a finite length, the analog value is determined by all bits in the window, where all bits have the same weight. The analog value of a Delta Sigma bit-stream can be simply negated by reversing the polarity of every bit in the stream. A few examples of first-order Delta Sigma-modulated (DSM) bit-streams are given in Table 1.

Table 1. Examples of first-order Delta Sigma-modulated (DSM) bit-streams. P, positive; N, negative.

Analog Value	Delta Sigma Stream								
1	P	P	P	P	P	P	P	P	...
0.5	P	P	P	N	P	P	P	N	...
0	P	N	P	N	P	N	P	N	...
−0.5	P	N	N	N	P	N	N	N	...
−1	N	N	N	N	N	N	N	N	...

The instantaneous analog value of a running Delta Sigma bit-stream can also be calculated using Equation (1) with a pre-determined window length, e.g., W bits. Thus, Equation (1) can be applied to both a bit sequence of finite length and a running bit-stream of infinite length. This result applies to any-order Delta Sigma modulated bit-streams, including first-order and higher-order Delta Sigma modulated bit-streams.

2.2. Techniques for Modifying Instantaneous Analog Values and/or the Length of the Delta Sigma-Modulated Bit-Streams

We present techniques for processing Delta Sigma bit-streams based on Equation (1). Assume that we have a Delta Sigma-modulated bit sequence of finite length. From Equation (1), we see that one P bit and one N bit in a Delta Sigma bit sequence cancel each other and do not contribute to the value of the numerator, but undoubtedly contributed to the length of the window, i.e., the value of the denominator. In other words, removing or adding P-N pairs in a Delta Sigma bit sequence changes its analog value by the ratio of the original length over the new length, as shown in Figure 4a. In another case, assume that m and M are integers and m is smaller than M. If we merge every M P bits and every M N bits, excluding P-N pairs, into m bits each and use P-N pairs to fill the space and keep the length of the bit sequence constant, then the analog value of the new bit-stream is reduced by m/M, as shown in Figure 4b. Moreover, if we merge every M P and every M N bits into m bits each and merge M P-N pairs from the original sequence into m P-N pairs, then the length of the Delta Sigma bit sequence is reduced by m/M, while its analog value is preserved, as shown in Figure 4c.

In summary, we have created three techniques to represent these operations, which are based on the P-N pair method:

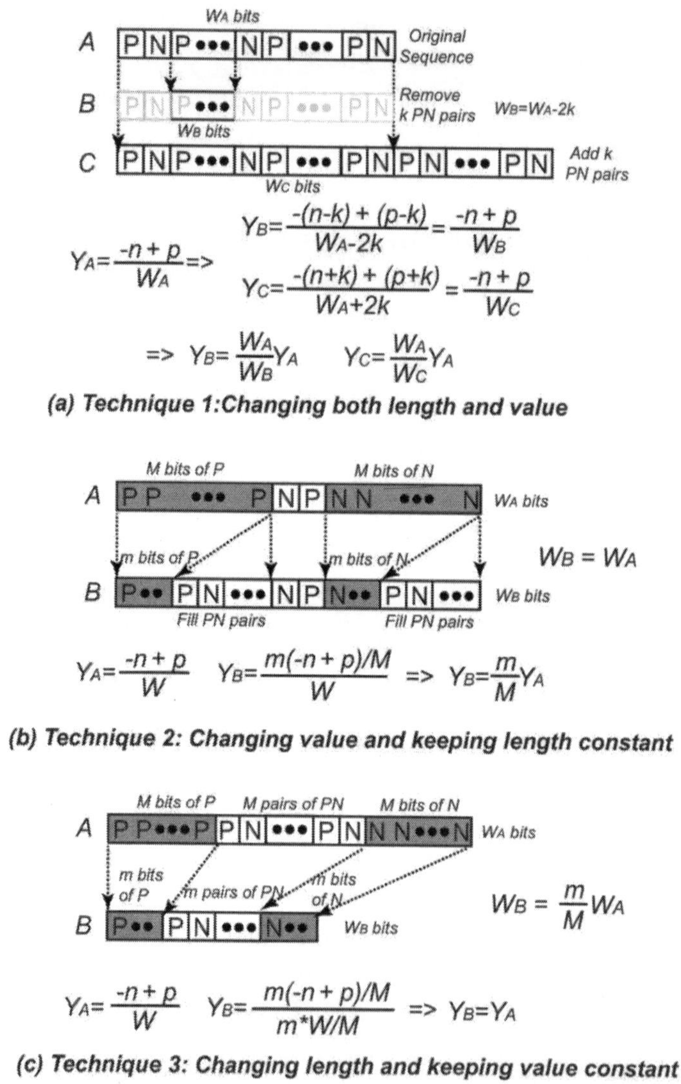

Figure 4. P-N pair method: Delta Sigma signal-processing techniques to adjust the length and value of a Delta Sigma bit sequence. A is the original Delta Sigma bit sequence with W bits, while B and C are the adjusted Delta Sigma bit sequence with lengths of W and W , respectively. (a) Changing the value by changing the length; (b) changing the value while keeping the length constant; and (c) changing the length while keeping the value constant.

Technique 1: Create a new Delta Sigma bit sequence by adding or removing P-N pairs from the original Delta Sigma bit sequence, changing its length from W bits to W bits. The new bit sequence has an analog value of W /W of the original bit sequence.

Technique 2: Create a new Delta Sigma bit sequence by merging every M homogeneous bits from the original Delta Sigma bit sequence into m bits and using P-N pairs to fill the space and keep the length of the sequence constant. The new Delta Sigma bit sequence has an analog value of m/M of the original sequence.

Technique 3: Create a new Delta Sigma bit sequence by merging every M homogeneous bits from the original Delta Sigma bit sequence into m bits and merging M P-N pairs into m P-N pairs. By doing so, the new sequence has the same analog value as the original sequence, while the length of the new bit sequence is reduced to m/M of its original length.

Our proposed Delta Sigma circuits are designed based on the above techniques.

3. DELTA SIGMA ADDERS

In this section, we introduce Delta Sigma adders using the proposed P-N pair method. Adding is a basic linear operation, which is indispensable in DSP. There are two types of adders in DSP: sum adders and average adders. In Delta Sigma processing, a sum adder generates a new Delta Sigma bit-stream based on the input bit-streams. The analog value of the output bit-stream is equal to the sum of the analog values of the addend bit-streams. On the other hand, an average adder generates a bit-stream whose analog value is equal to the average of all of the analog values of the addend bit-streams.

Here, we show a basic example of adding Delta Sigma bit-streams. An average bit-stream can be simply created by mixing all of the addend bit-streams bit by bit. As shown in Figure 5, the output bit sequence Z is generated by mixing input bit sequences A, B and C. Here, the analog value of Z is 1/3 of the sum (y + y + y). However, its length is three-times that of one addend. Therefore, Z requires a three-times faster clock than the addend bit-streams. In order to obtain the output bit-stream with the same data rate of the input bit-streams, the length of the mixed bit-stream must be reduced. This can be done using Technique 3 in Section 2 by deleting P-N pairs and homogeneous bits from the mixed bit sequence until the output bit sequence has the same length of the addends. This keeps the analog value while reducing the length of the bit sequence. On the other hand, by applying Technique 1 in Section 2, we can reduce the length of the bit sequence by three while increasing its analog value by three, thus removing the coefficient 1/3 and obtaining the sum bit sequence.

An on-the-fly processing circuit can be designed to perform the aforementioned processing of long combined bit-streams. The basic idea of an on-the-fly Delta Sigma adder is to remove P-N pairs simultaneously from the addend Delta Sigma bit-streams in order to keep the data rate of the sum bit-stream the same as that of the addends. This requires counting the number of P and N bits from all addends in each clock cycle. A counter performs the on-the-fly operation based on:

$$Y = \sum_{i=1}^{M} X_i \tag{2}$$

In Equation (2), Y is the counting result. X is the instantaneous bit of the i-th input bit-stream, and M is the number of addends. To make a P bit cancel an N bit, X equals +1 if the input bit is P, while X equals −1 if the input bit is N. This guarantees that P-N pairs can be removed from the result. Y represents the number of the surplus of either P bits or N bits. Thus, the result Y is either a zero, a positive number or a negative number. We use a signed binary number, for example the two's complement format, to represent the result Y.

Nevertheless, since +1 and −1 are applied in Equation (2), the system requires dual power supplies or a tri-state digital system similar to [21], which is not compatible with current two-state digital systems. We need a counter that can work in a two-state digital system. To make X compatible with the commonly-used binary system, we modify Equation (2) to get:

$$Y' = 2 \times \sum_{i=1}^{M} X_i - M \tag{3}$$

In Equation (3) X equals one if the input bit is P or equals zero if the input bit is N. A binary counter for calculating Equation (3) can be realized using an M-input one-bit binary adder in a two-state digital system. Here, a positive value of the Y' means Y' P bits remain after removing P-N pairs from all of the input bits, while a negative value of the Y' means −Y' N bits remain.

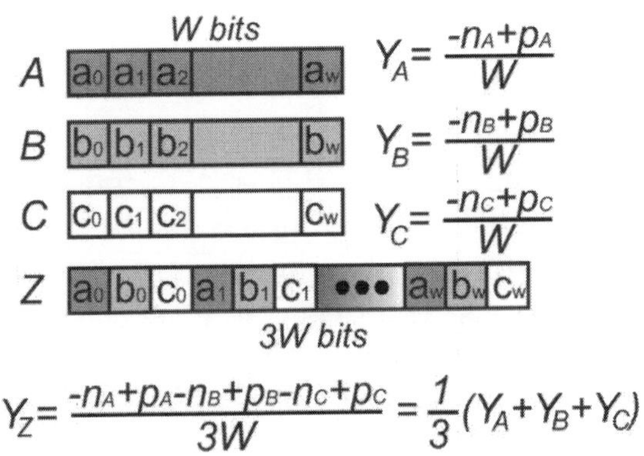

Figure 5. Averaging of the Delta Sigma bit-streams by mixing bits.

3.1. Delta Sigma Sum Adder

Based on Equation (3), we design a Delta Sigma sum adder. A sum adder takes M Delta Sigma bit-streams as inputs and generates a sum bit-stream. M is an arbitrary positive integer. In each clock cycle, the sum adder uses the counter in Equation (3) to count all of the input bits and obtains the result in a signed binary format. Besides the current result in the counter, the adder also needs to consider the previous value of the counter. An adder adds the previous value, which is stored in memory, to the current counter value. If the added result is positive (negative), then the sum output is one bit P (N). If the added result is greater than +1 or less than −1, then the added result is updated and stored in memory for the next clock cycle. The value stored in memory has the same polarity of the added result, but its absolute value is reduced by one from the added result. This is because the output takes one bit from the value. If the added result is zero, then the stored value is zero, and the output takes one bit from a toggling signal. This is because a toggling bit-stream has an analog value of zero, so adding toggling bits does not affect the analog value of the sum bit-stream.

A simplified schematic of a Delta Sigma sum adder is shown in Figure 6. X to X are Delta Sigma bit-streams as addends and inputs to a Counter. The Counter operates using Equation (3) and generates a signed binary number Y'. Y' is added to the value in Q by a binary adder. Here, Q is a register storing the previous results from the Counter. Q has the same number of bits as Y'. The added result of Y' and Q determines the output bit through C. Here, C is a digital comparator, whose output is one P bit if the input binary number is above zero or N if the input is less than zero. If the added result of Y' and Q is zero, then the output takes the value from a toggling bit T, and T toggles for the next cycle, i.e., T = NOT(T). If the added result of Y' and Q is not zero, a new value of Q is calculated and stored. The truth table of the circuit is summarized in Table 2. To avoid overflow errors, when Q reaches its maximum or minimum value, it should keep the value instead of rolling back to zero.

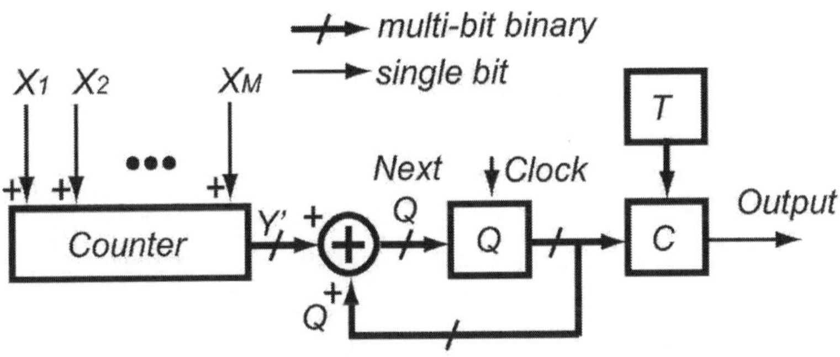

Figure 6. Simplified schematic of the Delta Sigma sum adder.

Table 2. Truth table of the sum adder.

Q + Y'	≥1	≤−1	0
Next Q	Q + Y' − 1	Q + Y' + 1	0
Output	P	N	T
T	T	T	NOT(T)

3.2. Delta Sigma Average Adder

Based on the Delta Sigma sum adder, we design the Delta Sigma average adder. The average adder generates an output Delta Sigma bit-stream, whose analog value is the average of all input Delta Sigma bit-streams. The average adder is essentially a digital Delta Sigma modulator, which can be realized in first-order or higher-order. The order number of the digital Delta Sigma modulator should be equal to or higher than the order number of the input bit-streams to maintain the signal-to-noise and distortion ratio (SNDR). For example, if the input bit-streams are generated by second-order Delta Sigma modulators, then the digital

Delta Sigma modulator in the average adder should be at least second-order. Otherwise, the SNDR of the output bit-stream decreases from that of the input.

Simplified schematics of the Delta Sigma average adders are shown in Figure 7. Figure 7a shows a first-order average adder, which has a similar structure to the sum adder in Figure 6. It also shares the same truth table with the sum adder. The difference is that the average adder has a feedback path from the output to the Counter. The equation of the Counter in an average adder is:

$$\begin{aligned} Y' &= 2 \times \sum_{i=1}^{M} X_i - M - M \times (2 \times Output - 1) \\ &= 2 \times \sum_{i=1}^{M} X_i - 2 \times M \times Output \end{aligned} \quad (4)$$

where Output equals one (zero) if it is P (N). Thus, in each clock cycle, the input bits are added to the feedback bits from the output. The maximum value of feedback is the same as the sum of the maximum value of all input bits. The feedback bits are negated from the output bit, i.e., every feedback bit is P if the output bit is N, or vice versa. The added result of Y' and Q determines the output bit and the next Q according to Table 2. We can also extend the first-order average adder to a second-order average adder, as shown in Figure 7b.

The average adder can also be used as a coefficient multiplier by a coefficient less than unity. The coefficient multiplier generates a new Delta Sigma bit-steam, whose analog value equals the analog value of the input bit-stream multiplied by a pre-defined coefficient. For example, the coefficient can be m/M, where both m and M are integers and m is smaller than M. This can be realized by using an M-input average adder with m of its inputs connected to the input bit-stream, while the other M – m inputs are connected to a toggling bit-stream, as shown in Figure 8a, where M = 4 and m = 3. The toggling bit-stream, which has an analog value of zero, is running at the same bit rate of the input bit-stream. When a higher resolution of the coefficient is required, a cascade structure can be applied using more stages of average adders, as shown in Figure 8b.

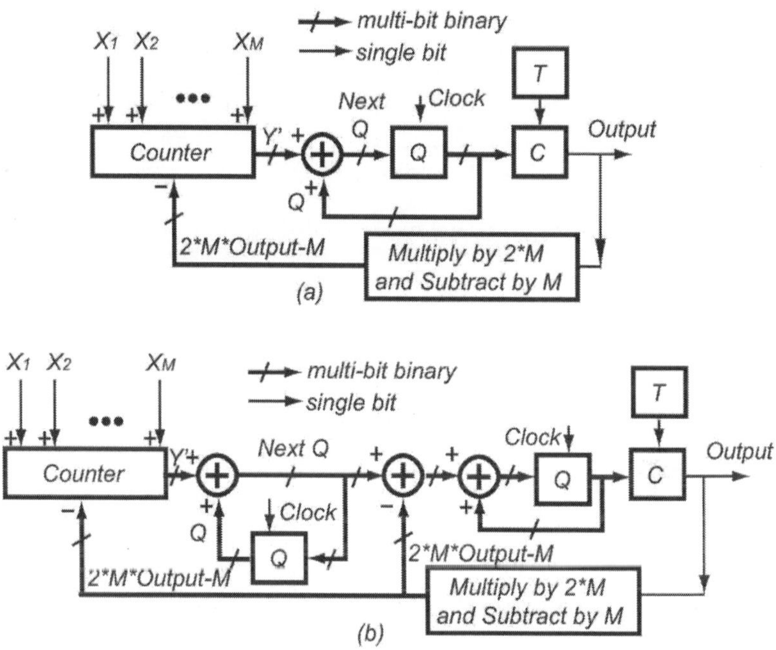

Figure 7. Simplified schematic of the Delta Sigma average adders. (a) First-order average adder; and (b) second-order average adder.

The proposed Delta Sigma adders have several advantages. First, they can take any number of input addend bit-streams compared to the two-input Delta Sigma adders in [22]. Second, the circuit complexity and power consumption are lower than other Delta Sigma adders. This is because when adding more addends, it only requires a linear increase in the size of the counter and memory. Therefore, the complexity of the proposed Delta Sigma adder circuits of N inputs is Θ(N), which is better than Θ(N) when circuits using sorting networks are applied [13]. Furthermore, compared to multi-bit binary adders, the proposed Delta Sigma adder can take addends at different clock rates, as long as the adder clock is equal to or faster than the highest clock rate among all addends.

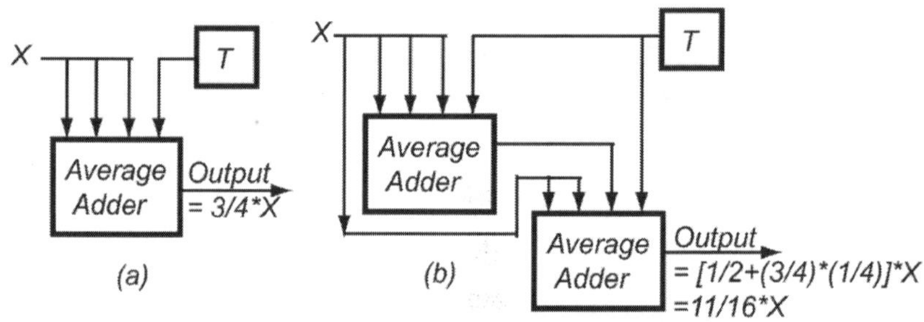

Figure 8. Simplified schematic of the Delta Sigma coefficient multiplier based on the Delta Sigma average adder: (a) one stage; and (b) cascade structure.

4. SIMULATION OF DELTA SIGMA ADDERS

This section presents simulation results of the Delta Sigma sum adder and the average adder. We simulated the proposed sum adder and the average adder in MATLAB using the Delta Sigma Toolbox [34]. Figure 9 shows the simulated SNDR plotted against the input power of a two-input average adder and a two-input sum adder both at an oversampling ratio (OSR) = 1024, where the inputs are Delta Sigma-modulated sine waves. Because the sum adder does not have feedback, the previous value Q representing the "temporal overflow" bits may not be able to be completely assimilated by the output bit-stream. The temporal overflow remains in the system until the input bit-stream changes the polarity of its analog value. Therefore, the temporal overflow can stay in the system for a long time and affects the output results. This process introduces extra noise in the output bit-stream. This noise increases when the input amplitude is high, because there are more "temporal overflow" bits at a higher amplitude. Furthermore, this noise increases with the number of addends. This is because when the number of addends is large, there are higher chances of having temporal overflow when input signals are in-phase. Figure 10 shows MATLAB-simulated SNDR as a function of OSR and the number of inputs in a Delta Sigma sum adder, where the input signals are sine waves modulated by second-order Delta Sigma modulators.

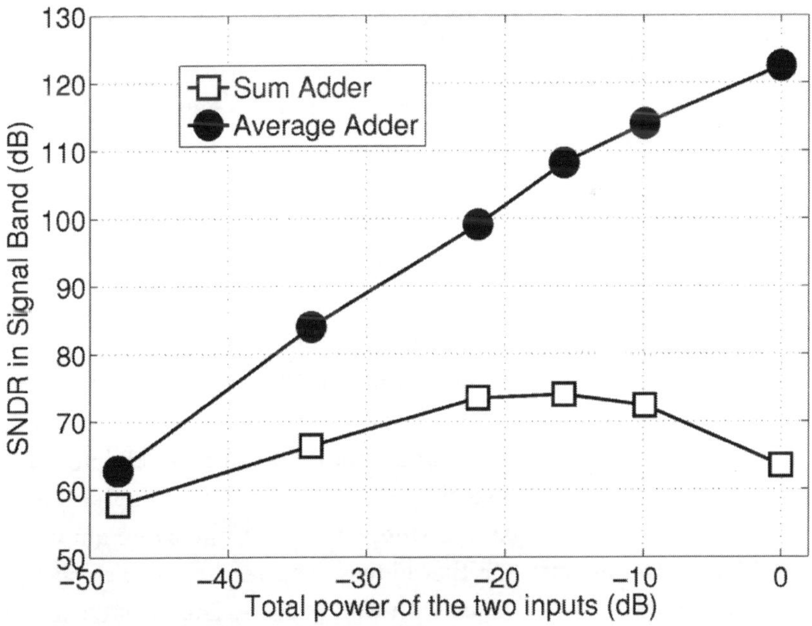

Figure 9. Simulated output signal-to-noise and distortion ratio (SNDR) v.s. input power in a two-input sum adder and a two-input second-order average adder both with oversampling ratio (OSR) = 1024.

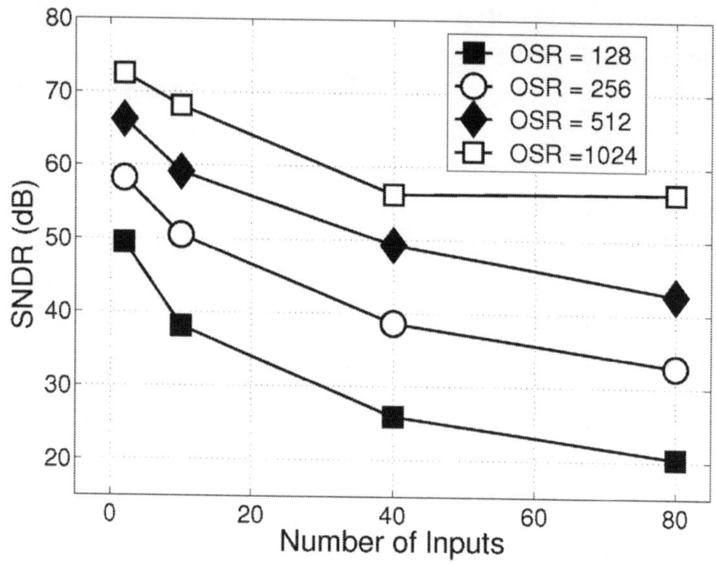

Figure 10. Simulated output SNDR in a Delta Sigma sum adder at different numbers of inputs and OSR.

We simulated the output bit-stream spectrums of the sum and average adders, as shown in Figure 11. Compared to the sum adder, the average adder has a feedback path that is able to completely merge its previous value with the current input bits within one clock cycle. Thus, there are no "temporal overflow" bits. Moreover, the feedback path enables the possibility of building higher-order digital Delta Sigma modulators that can further improve the noise shaping performance of the output bit-stream. In Figure 11, we use ten sine waves with the same amplitude and different frequencies for the adders. The ten inputs are second-order bit-streams of sine waves with OSR = 1024. The frequencies are normalized to the sampling frequency of the Delta Sigma modulators. Figure 11 shows that the noise floor of the sum adder is higher than the average adder by 50 dB. Furthermore, the noise shaping slopes in the sum adder and the first-order average adder are less than 30 dB/dec, while the output of the second-order average adder maintains a 40-

dB/dec noise shaping slope, which is the same as the spectrum of the ideal second-order sum bit-stream.

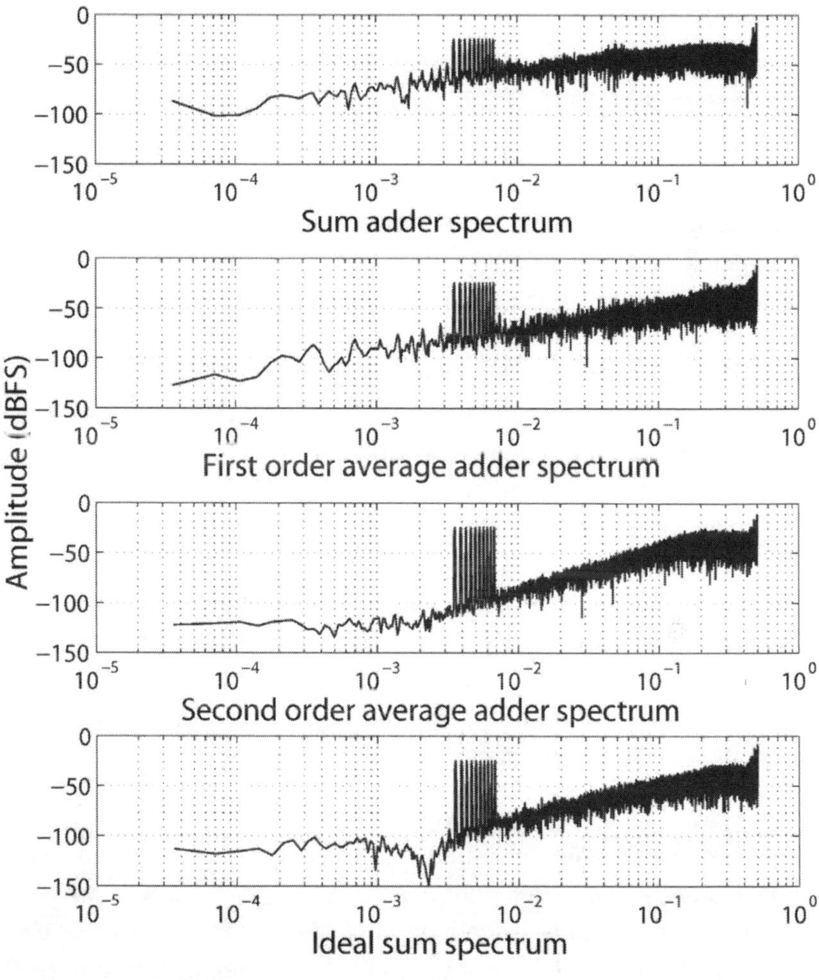

Figure 11. Simulated spectrums of Delta Sigma adders with ten addends.

In summary, while in a multi-bit binary DSP system, either a sum adder or an average adder can be used in digital filters, in a Delta Sigma DSP system, the average adder is preferred because of better noise performance.

5. DOWNSAMPLING AND DELTA SIGMA COMPRESSOR

In this section, we discuss the techniques of changing the length of a Delta Sigma bit sequence while preserving its analog value. This is equivalent to adjusting the clock rate of a Delta Sigma bit-stream. We are particularly interested in downsampling a Delta Sigma bit-stream. However, we cannot simply use a lower speed clock to sample the Delta Sigma bit-stream and hope to preserve its analog value. Downsampling of a Delta Sigma bit-stream is equivalent to data compression. An algorithm is necessary to preserve the analog value while reducing the number of bits. Unlike complicated compressing algorithms in the multi-bit system, this work proposes a simple method based on Technique 3 in Section 2. Essentially, we reduce the number of homogeneous bits and P-N pairs with the same ratio, for instance K. Then, the new bit sequence has the same analog value as the original bit sequence, and the length of the new bit sequence is 1/K of the original bit sequence. This algorithm can be realized with simple circuitry using on-the-fly processing.

Here, we give an example of a Delta Sigma compressor circuit to reduce the data rate by 1/2. The circuit is shown in Figure 12. Assume that a Delta Sigma bit-stream is coming into the compressor at a clock frequency f. We use an f/2 clock to sample two incoming bits at once. By doing so, any bit is only checked once in a two-bit window. We mark the two bits in the window as a and a , where a is the preceding bit, while a is the succeeding bit. If the two bits are homogenous, then the output bit X should be the same polarity of the inputs. If the two incoming bits are heterogeneous, we take the value from a toggling bit T as the output bit.

The toggling bit T toggles whenever the incoming two bits are heterogeneous. Thus, both the number of P-N pairs and the number of homogeneous bits are reduced by 1/2 at the output. The value of X at frequency f/2 is the output Delta Sigma bit-stream. The Delta Sigma compressor can also be realized using the Delta Sigma average adder by adding sub-sequences from the original sequence, at the cost of increase of circuit complexity.

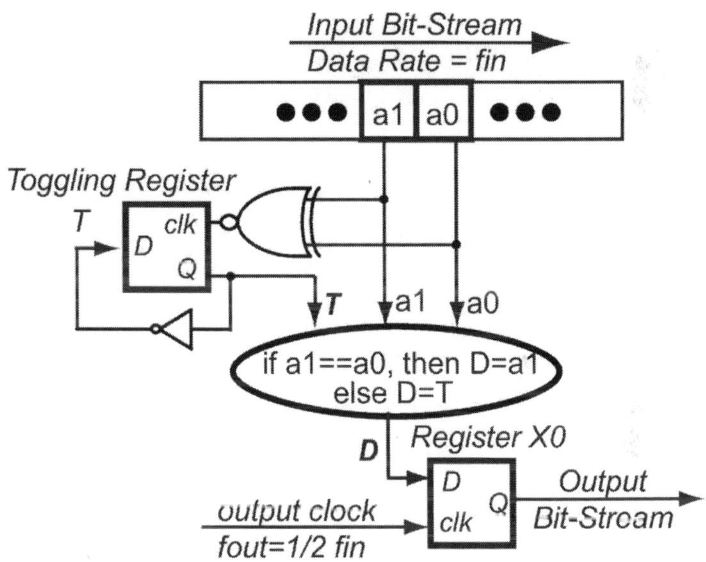

Figure 12. Simplified circuit schematic of the Delta Sigma compressor.

We simulated the SNDR of the compressor in MATLAB. Figure 13 shows the SNDR of the input and output of the Delta Sigma compressor versus OSR. The input is created by a second-order Delta Sigma modulator. Although the algorithm keeps the output analog value the same as the input analog value, Figure 13 shows that the compressor reduces the SNDR because the OSR of the output bit-stream is half that of the input bit-stream. Figure 14 shows the simulated input and output spectra of the Delta Sigma compressor with a second-order input bit-stream and

OSR = 1024. The noise floor of the output spectrum increased by 50 dB compared to the input spectrum. Furthermore, the noise shaping slope of the output spectrum reduced to 20 dB/dec, which makes the output a first-order bit-stream. The compressor based on the first-order sum adder can only preserve the noise shaping of a first-order Delta Sigma bit-stream.

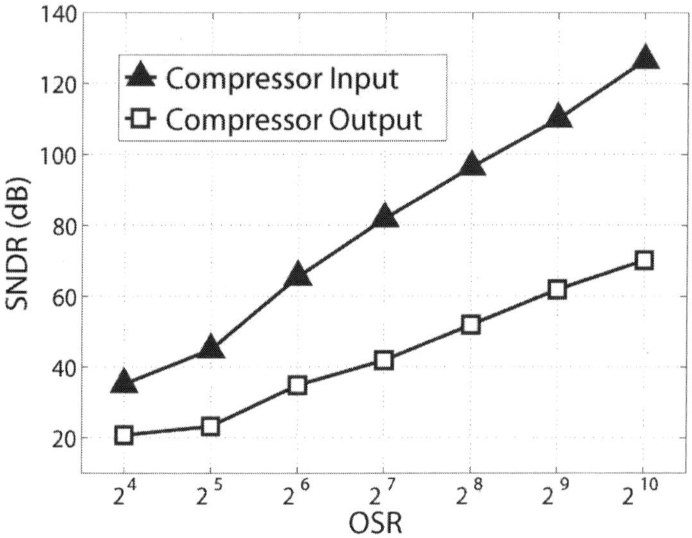

Figure 13. Simulated signal-to-noise-ratio of the Delta Sigma compressor.

6. MEASUREMENT RESULTS

We tested the proposed Delta Sigma signal processing circuits using a field programmable gate array (FPGA). The testbed is shown in Figure 15. We used second-order Delta Sigma modulators AD7401 [35] to generate the Delta Sigma bit-streams from analog signal generators. The Delta Sigma bit-streams are fed into the proposed Delta Sigma processing circuits, which include the sum adder and the average adder. The processing circuits are coded in Verilog and simulated by the Xilinx ISE isim simulator and then implemented on an Opal Kelly Spartan-3 FPGA board [36]. In our experiment, an external function generator provided the clock signals to both the Delta Sigma modulator and the Delta Sigma processing circuit. Fourth-order analog filters recovered analog waveforms from both the Delta Sigma modulator and the Delta Sigma processing circuit. Oscilloscopes calculated the spectra of the digital signals using the FFT function. Both the AD7401's output port and the Opal Kelly FPGA are powered by a 3.3-V power supply.

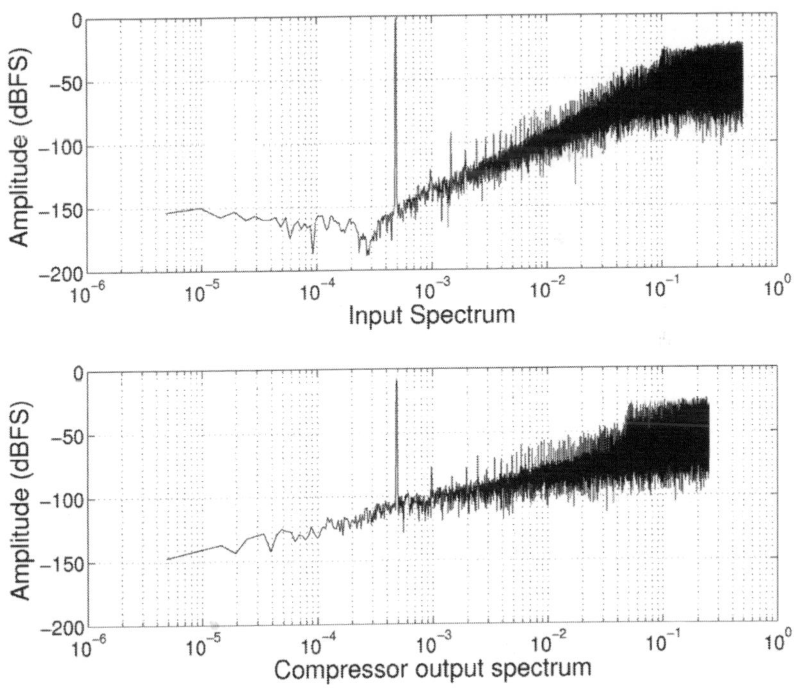

Figure 14. Simulated input and output spectrum of the Delta Sigma compressor.

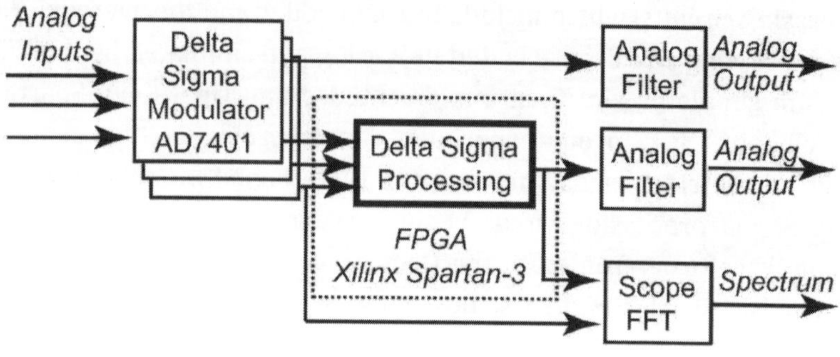

Figure 15. Testbed setup for testing the Delta Sigma processing circuits.

Figure 16 shows the input and output waveforms of a three-input second-order average adder. The three input signals are 113-Hz, 157-Hz and 213-Hz sine waves all with a 200-mVpp amplitude. The output waveform is recovered from a fourth-order analog filter. The output amplitude is higher than the amplitude of the inputs because of the internal gain of the AD7401. Figure 17 shows the FFT result of the second-order average adder output, which is calculated by the oscilloscope. In Figure 16 and Figure 17, the functionality of the three-input average adder is verified. Furthermore, the 40-dB/dec noise shaping slope in Figure 17 shows that the processed signal is a second-order Delta Sigma bit-stream. In order to get a better view of the noise shaping slope, the OSR is set at approximately 100 for the three signals, which makes the sampling clock of the AD7401 and the clock of the Delta Sigma average adder both at 4 kHz.

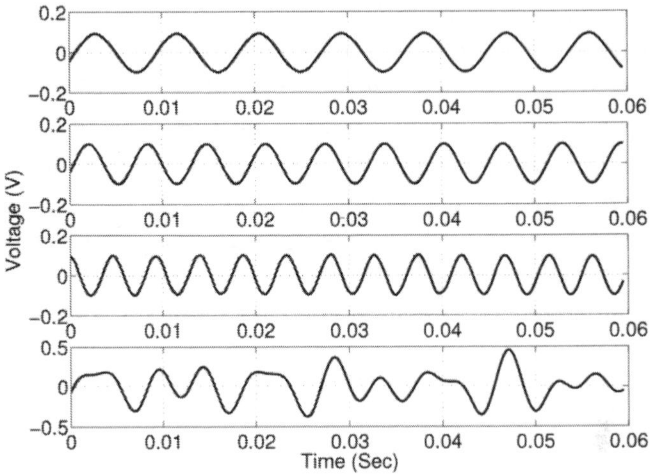

Figure 16. Measured time-domain waveforms of a three-input Delta Sigma second-order average adder recovered by analog filters. From top to bottom: Input 1, Input 2, Input 3 and output.

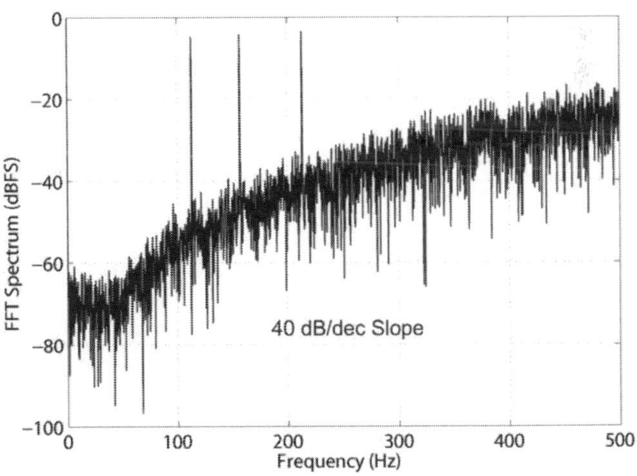

Figure 17. Measured FFT spectrum of the output of a three-input Delta Sigma second-order average adder. The noise shaping slope is 40 dB/dec.

We also recorded the digital input and output waveforms of the Delta Sigma compressor, as shown in Figure 18. In the test, we give a DC input of 500 mV and use a sampling rate of 500 kHz to obtain the input second-order Delta Sigma bit-stream. The output of the Delta Sigma compressor has half the data rate of the input Delta Sigma bit-stream. The input and output bit-streams have the same average duty cycle, which means the analog value of the input is preserved in the output.

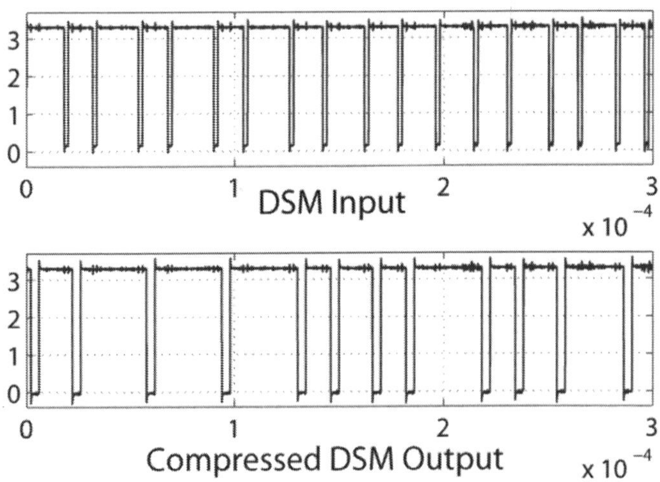

Figure 18. Measured time-domain input and output waveforms of the Delta Sigma compressor.

7. COMPARISON BETWEEN DELTA SIGMA ADDERS AND MULTI-BIT ADDERS

In linear DSP systems, higher-order FIR filters require adding many addends together at once. The proposed Delta Sigma processing circuits can significantly reduce circuit complexity compared to multi-bit processing circuits, especially when the number of inputs in an adder is large. In Figure 19, we tested the integrated circuit area of the Delta

Sigma sum adder, average adder and a multi-bit ripple-carry adder. The Delta Sigma sum adder has a simpler structure than the average adder. However, since the sum adder introduces extra noise, it is fair to compare the multi-bit binary adder and the Delta Sigma average adder, since neither circuit adds noise. We synthesized the Verilog code in the IBM 0.18-μm CMOS process using the Cadence RTL Compiler. The results show that if we replace the binary adder with the Delta Sigma average adder, the circuit area is reduced by 94% when the number of inputs is 100.

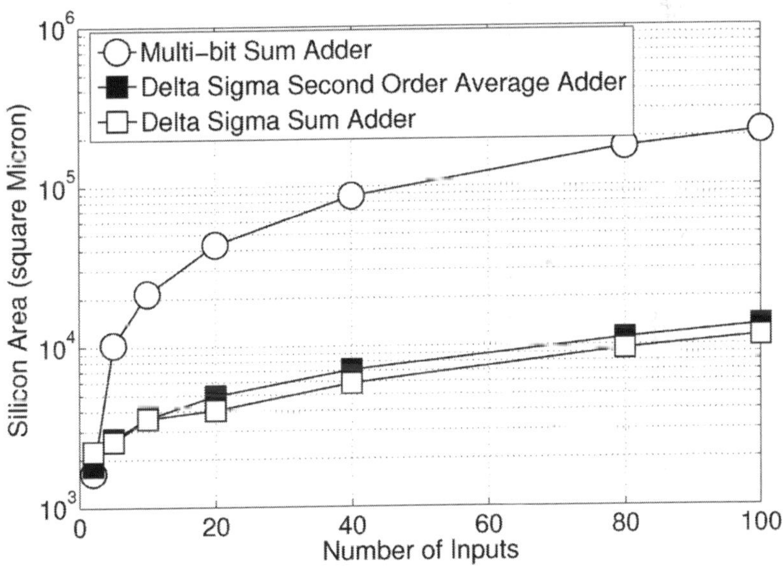

Figure 19. Comparison of silicon area between the Delta Sigma adders and the binary adder.

We also compared the power consumption between the proposed Delta Sigma adders and the multi-bit adder. In order to make a fair comparison, we refer to Texas Instruments ADS131E04 [14]. It contains a Delta Sigma ADC made of a second-order Delta Sigma modulator, which operates at a fixed clock of 1024 kHz. The output sample rate of the ADC is 16 kSamples/s with 24-bit resolution. This means that, if we use the Delta Sigma modulator in the ADS131E04 as the ADC and make the Delta Sigma sampling rate 1024 kSamples/s, the result is equivalent to having a 24-bit binary ADC with a sampling rate of 16 kSamples/s. Thus, utilizing the same resolution and sampling rate of the ADC, when we design DSP circuits, it is fair to compare the power consumption of the proposed Delta Sigma average adder at a 1024-kHz clock rate to a 24-bit adder at a 16-kHz clock rate. We compared the power consumption for different numbers of inputs between the Delta Sigma adders and multi-bit binary adders. The comparison includes leakage power, net power and switching power, which are shown in Figure 20, Figure 21 and Figure 22, respectively. For a 100-input adder, the proposed Delta Sigma average adder reduces the switching power by more than 96% compared to the multi-bit adder.

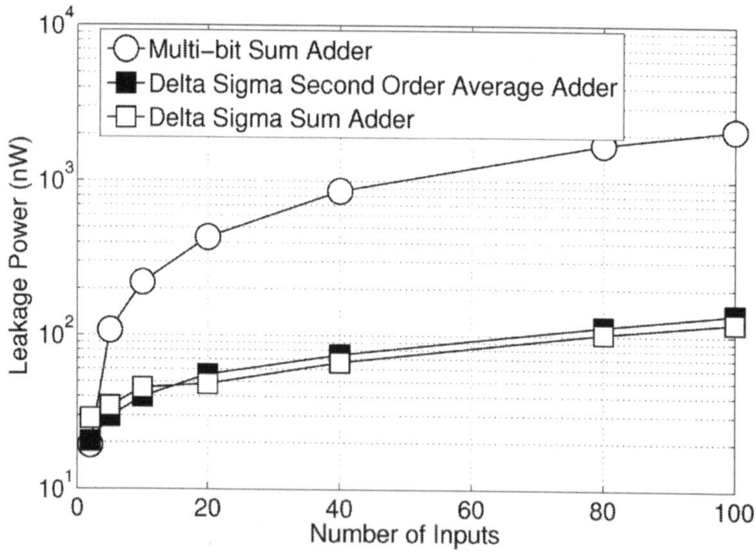

Figure 20. Comparison of leakage power between the Delta Sigma adders and a binary adder.

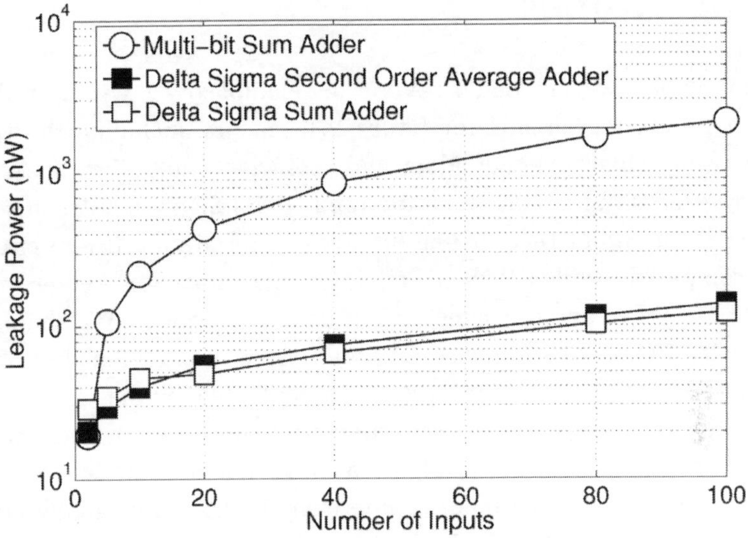

Figure 21. Comparison of net power between the Delta Sigma adders and a binary adders.

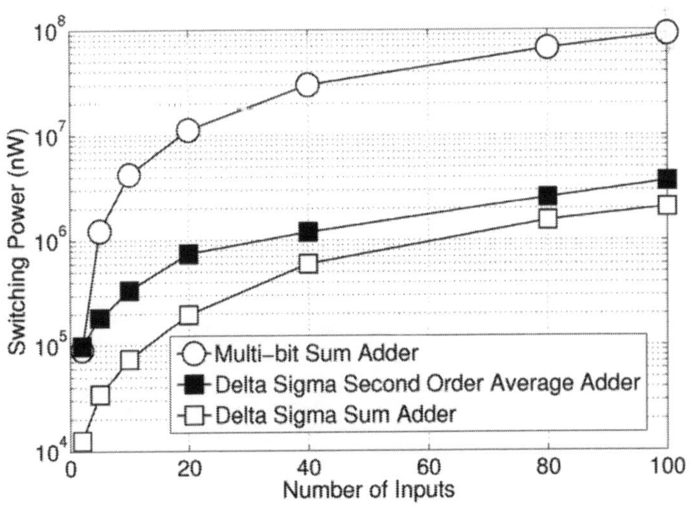

Figure 22. Comparison of switching power between the Delta Sigma adders and a binary adder.

8. CONCLUSIONS

In this paper, we presented several low-complexity low-power Delta Sigma processing circuits for future IoT sensors, including the Delta Sigma sum adder, average adder and coefficient multiplier. These on-the-fly processing circuits take the input Delta Sigma bit-streams and generate an output Delta Sigma bit-stream in real time. These circuits are compatible with CMOS digital logic. A comparison between our proposed circuit and other methods is shown in Table 3. We simulated the Delta Sigma adders in MATLAB and tested the functionality of the proposed circuits on an FPGA. We also presented synthesis results in the IBM 0.18-μm process. We compared silicon area and power consumption between Delta Sigma average adders and conventional multi-bit binary adders. In our comparison of Delta Sigma and multi-bit adders, we chose 100 addends. FIR (finite impulse response) filters with 100 or more taps are not uncommon in DSP algorithms, where each tap represents one multiplication and one addend in the physical realization of the filter. Synthesis results showed that the proposed Delta Sigma adders can significantly reduce silicon area and power consumption, which makes them competitive candidates for future low-power, low-complexity DSP circuits for IoT.

Table 3. Comparison between Circuits.

Comparison	[22] **1978**	[21] **2009**	[13] **2012**	**This Work 2015**
Circuit Structure	Binary Adder	Tri-State Digital	Sorting Network	Counter
DSM Order	First-Order Only	First-Order Only	First- and Higher Order	First- and Higher Order
Circuit Complexity	$\Theta(N)$	$\Theta(N)$	$\Theta(N\log(N^2))$	$\Theta(N))$

AUTHOR CONTRIBUTIONS

Wei Tang conceived of the idea and directed the work. Yifei Liu performed the simulation, the experiment, the synthesis and obtained the data. Wei Tang wrote the manuscript. Paul M. Furth helped with earlier versions of the manuscript.

REFERENCES

1. Sai, V.; Mickle, M. Exploring energy efficient architectures in passive wireless nodes for IoT applications. IEEE Circuits Syst. Mag. 2014, 14, 48–54.

2. Bachmann, C.; Ashouei, M.; Pop, V.; Vidojkovic, M.; Groot, H.; Gyselinckx, B. Low-power wireless sensor nodes for ubiquitous long-term biomedical signal monitoring. IEEE Commun. Mag. 2012, 50, 20–27.

3. Chen, G.; Hanson, S.; Blaauw, D.; Sylvester, D. Circuit design advances for wireless sensing applications. Proc. IEEE 2010, 98, 1808–1827.

4. Fiscal Year 2014 Historical Tables Budget of the U.S. Government; Office of Management and Budget: Alexandria, VA, USA, 2014.

5. Yoo, J.; Yan, L.; El-Damak, D.; Altaf, M.; Shoeb, A.; Chandrakasan, A. An 8-channel scalable EEG acquisition SoC with patient-specific seizure classification and recording processor. IEEE J. Solid-State Circuits 2013, 48, 214–228.

6. Yoo, J.; Yan, L.; El-Damak, D.; Bin Altaf, M.; Shoeb, A.; Yoo, H.J.; Chandrakasan, A. An 8-channel scalable EEG acquisition SoC with fully integrated patient-specific seizure classification and recording processor. In Proceedings of the 2012 IEEE International Solid-State Circuits Conference Digest of Technical Papers (ISSCC), San Francisco, CA, USA, 19–23 February 2012; pp. 292–294.

7. Qian, C.; Shi, J.; Parramon, J.; Sanchez-Sinencio, E. A low-power configurable neural recording system for epileptic seizure detection. IEEE Trans. Biomed. Circuits Syst. 2013, 7, 499–512.

8. Custodio, J.; Goes, J.; Paulino, N.; Oliveira, J.; Bruun, E. A 1.2-V 165-μW 0.29-mm2 multibit sigma-delta ADC for hearing aids using nonlinear DACs and with over 91 dB dynamic-range. IEEE Trans. Biomed. Circuits Syst. 2013, 7, 376–385.

9. Porrazzo, S.; Morgado, A.; Segundo Bello, D.; Cannillo, F.; van Hoof, C.; Yazicioglu, A.; van Roermund, A.; Cantatore, E. A 155 μW 88-dB DR discrete-time ΔΣ modulator for digital hearing aids exploiting a summing SAR ADC quantizer. IEEE Trans. Biomed. Circuits Syst. 2013, 7, 573–582.

10. Deligoz, I.; Naqvi, S.; Copani, T.; Kiaei, S.; Bakkaloglu, B.; Je, S.S.; Chae, J. A MEMS-based power-scalable hearing aid analog front end. IEEE Trans. Biomed. Circuits Syst. 2011, 5, 201–213.

11. Lopez-Morillo, E.; Carvajal, R.; Munoz, F.; El Gmili, H.; Lopez-Martin, A.; Ramirez-Angulo, J.; Rodriguez-Villegas, E. A 1.2-V 140-nW 10-bit sigma-delta modulator for electroencephalogram applications. IEEE Trans. Biomed. Circuits Syst. 2008, 2, 223–230.

12. Stanacevic, M.; Murari, K.; Rege, A.; Cauwenberghs, G.; Thakor, N. VLSI potentiostat array with oversampling gain modulation for wide-range neurotransmitter sensing. IEEE Trans. Biomed. Circuits Syst. 2007, 1, 63–72.

13. Fujisaka, H.; Kamio, T.; Ahn, C.J.; Sakamoto, M.; Haeiwa, K. Sorter-based arithmetic circuits for sigma-delta domain signal processing—Part I: Addition, approximate transcendental functions, and log-domain operations. IEEE Trans. Biomed. Circuits Syst. 2012, 59, 1952–1965.

14. ADS131E04 Analog Front-End for Power Monitoring, Control and Protection; Technical Report; Texas Instruments: Harris, TX, USA, 2013.

15. Melanson, J. Signal Processing System Using Delta-Sigma Modulation Having An Internal Stabilizer Path With Direct Output-to-Integrator Connection. U.S. Patent 7,696,913, 13 April 2010. Available online: https://www.google.com/patents/US7696913.

16. Choi, J.; Han, S.; Jang, J.; Yu, H. Structure of Delta-Sigma Fractional Divider. U.S. Patent App. 10/179,840, 12 June 2003. Available online: https://www.google.com/patents/US20030108143.

17. Cho, M.; Kim, Y.; Kwon, J. Coefficient Multiplier and Digital Delta-Sigma Modulator Using the Same. U.S. Patent App. 12/783,294, 16 June 2011. Available online: https://www.google.com/ patents/ US20110140940.

18. Niwa, A.; Ueno, Y. Delta-Sigma Modulator and Signal Processing System. U.S. Patent 8,581,763, 12 November 2013. Available online: https://www.google.com/patents/US8581763.

19. Moue, T. Delta-Sigma Modulator and Signal Processing System. U.S. Patent 8,581,764, 12 November 2013. Available online: https:// www.google.com/patents/US8581764.

20. Dias, V.F. Signal processing in the sigma-delta domain. Microelectron. J. 1995, 26, 543–562.

21. Ng, C.W.; Wong, N.; Ng, T.S. Bit-stream adders and multipliers for tri-level sigma delta modulators. IEEE Trans. Circuits Syst. Express Briefs 2007, 54, 1082–1086. [Google Scholar]

22. Kouvaras, N. Operations on delta-modulated signals and their application in the realization of digital filters. Radio Electron. Eng. 1978, 48, 431–438.

23. Kouvaras, N. Some novel elements for delta-modulated signal processing. Radio Electron. Eng. 1981, 51, 241–250.

24. Kouvaras, N.; Karakatsanis, J. A technique for a substantial reduction of the quantization noise in the direct processing of delta-modulated signals. Signal Process. 1985, 8, 107–119.

25. Kouvaras, N. Modular network for direct complete addition of delta-modulated signals with minimum quantization error. Int. J. Electron. 1985, 59, 587–595.

26. Hidaka, Y.; Fujisaka, H.; Sakamoto, M.; Morisue, M. Piecewise linear operations on sigma-delta modulated signals. In Proceedings of the 9th International Conference on Electronics, Circuits and Systems, Dubrovnik, Croatia, 15–18 September 2002; Volume 3, pp. 983–986.

27. Zrilic, D. Method and Apparatus for Mixed Analog and Digital Processing Of Delta Modulated Pulse Streams Including Digital-to-Analog Conversion of A Digital Input Signal. U.S. Patent 5,349,353, 20 September 1994. Available online: https://www.google.com/patents/US5349353.

28. Circuits and Methods for Functional Processing of Delta Modulated Pulse Density Stream. U.S. Patent 6,285,306, 4 September 2001. Available online: https://www.google.com/patents/US6285306.

29. Horianopoulos, S.; Anastassopoulos, V.; Deliyannis, T. Design technique for hardware reduction in delta modulation fir filters. Int. J. Electron. 1991, 71, 93–106. [Google Scholar] [CrossRef]

30. Pneumatikakis, A.; Anastassopoulos, V.; Deliyannis, T. Realization of a high-order iir delta sigma filter. Int. J. Electron. 1995, 78, 1071–1089.

31. O'Leary, P.; Maloberti, F. Bit stream adder for oversampling coded data. Electron. Lett. 1990, 26, 1708–1709.

32. Liang, Y.; Wang, Z.; Meng, Q.; Guo, X. Design of high speed high snr bit-stream adder based on delta sigma modulation. Electron. Lett. 2010, 46, 752–753.

33. McGrath, D. Method and Apparatus for Processing Sigma-Delta Modulated Signals. U.S. Patent 5,990,818, 23 November 1999. Available online: https://www.google.com/patents/US5990818.

34. Schreier, R. Delta Sigma Toolbox: High-Level Design and Simulation Of Delta-Sigma Modulators. Matlab Toolbox. Available online: http://www.mathworks.com/matlabcentral/fileexchange/19-delta-sigma-toolbox (accessed on 29 December 2011).

35. AD7401 Isolated Sigma-Delta Modular; Technical Report; Analog Device Inc.: Norwood, MA, USA, 2011.

36. Opal Kelly XEM 3001v2 User's Manual; Technical Report; Opal Kelly Inc.: Portland, OR, USA, 2004.

CHAPTER 5

Noise Removal in Speech Processing Using Spectral Subtraction

Marc Karam[1], Hasan F. Khazaal[2], Heshmat Aglan[3], Cliston Cole[1]

[1]Department of Electrical Engineering, Tuskegee University, Tuskegee, USA

[2]Department of Electrical Engineering, Wasit University, Wasit, Iraq

[3]Department of Mechanical Engineering, Tuskegee University, Tuskegee, USA

ABSTRACT

Spectral subtraction is used in this research as a method to remove noise from noisy speech signals in the frequency domain. This method consists of computing the spectrum of the noisy speech using the Fast Fourier Transform (FFT) and subtracting the average magnitude of the noise spectrum from the noisy speech spectrum. We applied spectral subtraction to the speech signal "Real graph". A digital audio recorder system embedded in a personal computer was used to sample the speech signal "Real graph" to which we digitally added vacuum cleaner noise. The noise removal algorithm was implemented using Matlab software by storing the noisy speech data into Hanning time-widowed half-overlapped data buffers, computing the corresponding spectrums using the FFT, removing the noise from the noisy speech, and reconstructing the speech back into the time domain using the inverse Fast Fourier Transform (IFFT). The performance of the algorithm was evaluated by calculating the Speech to Noise Ratio (SNR). Frame averaging was introduced as an optional technique that could improve the SNR. Seventeen different configurations with various lengths of the Hanning time windows, various degrees of data buffers overlapping, and various numbers of frames to be averaged were investigated in view of improving the SNR. Results showed that using one-fourth overlapped data buffers with 128 points Hanning windows and no frames averaging leads to the best performance in removing noise from the noisy speech.

KEYWORDS

Speech Processing, Spectral Subtraction, Noise Removal, Fast Fourier Transform, Inverse Fast Fourier Transform

1. INTRODUCTION

Speech communications are used daily in our lives. Every case of speech communication involves a speaker, a listener, and various communication devices. Speech communications take place everywhere, such as domestic homes, work, school conferences, seminars, medical appointments, and cocktail parties. Often, random noises corrupt the communication between the speaker and the listener. These noises can cause speech to be heard incorrectly. Noises exist everywhere, and are produced by many factors, so that it is impossible to identify them all. The characteristics of these noises are either known or unknown; however, they all can distort, disrupt, or disguise the quality of speech signals. Therefore, background noises and noisy environments are likely to affect many people, especially people with hearing loss. The area of research that investigates removing noise from corrupted speech utilizing various signal processing methods is called speech processing. There are many different forms of speech processing such as speech enhancement, speech recognition, speech coding, and speech synthesis.

In recent studies, numerous filter designs have been implemented in communication systems to reduce and eventually eliminate the effects of incoming background noise, as well as to enhance speech intelligibility [1] -[5] . Removal of high frequency noise for speech enhancement using Frequency Response Masking (FRM), a technique based on designing low complexity, narrow transition bandwidth, linear phase Finite Impulse Response (FIR) filters, has been implemented [1] . An FIR filter has been designed to have impulse responses associated with various cut-off frequencies leading to a decrease in the Mean Square Error (MSE) when comparing original and filtered speech signals [2] . In applications where both the speech and the noise signals change continuously, adaptive filtering based on using three algorithms: Least Mean Square (LMS), Normalized Least Mean Square (NLMS), and Sign-Data Least Mean Square (SDLMS) algorithms has been implemented [3] . Discrete Wavelet Transform (DWT) algorithm was used for speech signal denoising with both hard and soft thresholding, with soft thresholding performing better than hard thresholding at all input SNR levels [4] . Residual musical noise resulting from spectral subtraction technique has been reduced using scaling factors and weighted functions [5] .

In this research, we focused on spectral subtraction noise removal approach in speech processing [6] . Our experiment involved sampling two different signals: a real-time speech signal "Real graph" and a noise signal generated by a vacuum cleaner. Using Matlab, we digitally added the vacuum cleaner noise to the speech signal "Real graph", thus obtaining a noisy speech signal. Noise removal cannot be successfully implemented in the time domain; rather, it is performed in the frequency domain. Our spectral subtraction noise removal approach involves segmenting the noisy speech signal into half-overlapped time domain data buffers multiplied by a Hanning window and then transforming the result into the frequency domain using the fast Fourier transform (FFT). Subsequently, noise is removed by subtracting the average magnitude of the noise spectrum from the noisy speech spectrum and zeroing out the negative values using half-wave rectification. Finally, after removing the noise from the noisy speech, we reconstructed the noise-reduced speech back to time domain using the Inverse Fast Fourier Transform (IFFT) [7] . We were able to listen to the reconstructed speech and we observed that the noise had effectively been reduced. Statistical evaluation of the results was accomplished by calculating the Speech to Noise Ratio (SNR) [8] . In order to improve the performance, we applied the technique of frames averaging [9] . Moreover, we studied the effect of varying the overlapping lengths of the data buffers and the Hanning windows on improving the SNR.

2. TIME DOMAIN TO FREQUENCY DOMAIN CONVERSION USING FFT

2.1. Sampling of the Noisy Speech "Real Graph" by Using the A-to-D Converter

First, consider the clean speech "Real graph" S(t) corrupted by vacuum cleaner noise N(t) and shown in Figure 1. The noisy speech "Real graph" is a continuous-time function which is converted to an electrical signal X(t) using a microphone connected to a digital audio recorder system. The microphone performs this conversion by detecting the changing air pressure of the audio sound. The electrical signals are transmitted through a cable wire or a median that is connected between the microphone and the digital audio recorder system.

The digital audio recorder system is an example of an Analog-to-Digital (A-to-D) converter. The A-to-D converter transforms the continuous-time noisy speech into a discrete-time noisy speech X[n]. A discrete-time signal is a non-continuous time signal. It has been sampled from a continuous-time signal using a digital audio recorder system. Discrete-time signals symbolize an indexed sequence of discrete-time samples. A continuoustime signal is sampled at equally spaced time impulses $t_n = nT_s$ as follows

$$X[n] = X(nT_s), \quad (1)$$

where T_s is the sampling period or fixed time between each sample. Each impulse value of X[n] is called sample of the discrete-time signal. The sampling period can also be represented as a fixed sampling rate:

$$f_s = \frac{1}{T_s} Hz \quad . (2)$$

In this research, a clean speech "Real graph" S[n] was recorded using software called sound recorder that was installed on a personal computer. A time waveform of the speech "Real graph" is shown in Figure 2(a).

The speech was recorded for a duration of 645 ms. Shannon sampling theorem states that any continuous-time signal with maximum spectrum frequency f_{max} can be reconstructed exactly from its samples $X[n] = X(nT_s)$ if the samples are taken at a sampling rate that is greater than 2 f_{max}. Since audio frequencies of audible sounds range from 20 to 20,000 Hz, thus, in our application, f_{max} is approximately 20 kHz. The sampling rate was automatically computed by Matlab, and had the value 44.1 kHz, which is bigger than twice 20 kHz, and thus satisfies the Shannon sampling theorem. There are a total of 28,446 samples with time space interval of 22.67 μs between each sample. A time waveform of a vacuum cleaner noise N[n] was also sampled for 645 ms at a rate of 44.1 kHz and is shown in Figure 2(b). The vacuum cleaner noise was digitally added to the clean speech. The sum of the two signals generates the noisy speech signal "Real graph" X[n] shown in Figure 2(c).

2.2. Storing the Noisy Speech "Real Graph" Using the Half-Overlapped Data Buffers

The noisy speech is the data we want to evaluate for noise removal. Once Matlab retrieves, reads, and formats the data in numerical value, the data are stored into segments. Each segment contains 256 samples of the noisy speech. Each segment is called a data-buffer $\hat{X}[n]$. Each data buffer half-overlaps another data buffer by a total of 128 samples. Our noisy speech "Real graph" has 221 half-overlapped data buffers that cover the entire length of the noisy speech data.

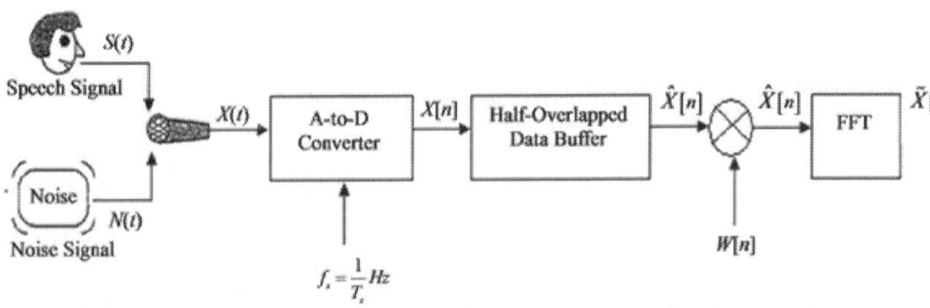

Figure 1. Block diagram of noisy speech generation and discretization.

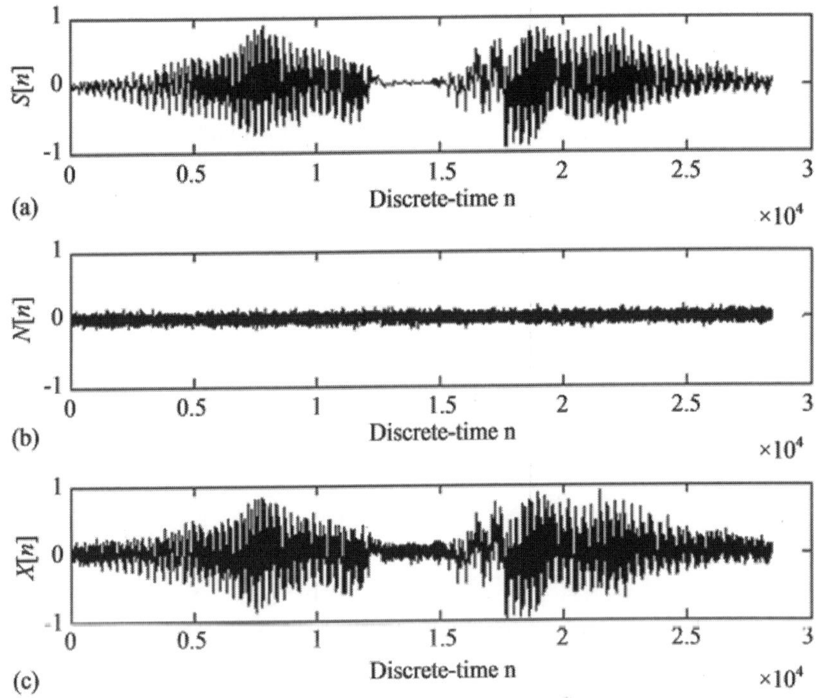

Figure 2. (a) Clean speech "Real graph" signal S(n); (b) Vacuum cleaner noise signal N(n); (c) Noisy speech "Real graph" corrupted by the vacuum cleaner noise X(n).

2.3. Analyzing the Noisy Speech "Real Graph" Using the Hanning Time Window

The noisy speech data of the 221 half-overlapped data buffers contain 28,446 samples; the transformation of the 28,446 samples from time domain to frequency domain using FFT would take a very long time for the computer to process and compute the spectrum. Computing a smaller amount of data at a time optimizes the efficiency of the computer processing speed. In this research, the noisy speech data of the 221 half-overlapped data buffers were decomposed into time windows called Hanning time windows. The Hanning time window is a bell curve shape that multiplies the noisy speech data of the half-overlapped data buffers. The portions of the noisy speech data that lie outside the Hanning time window are zeroed-out, while the portions inside are further evaluated for processing. The mathematical general expression of a Hanning time window is in the form

$$W[n]=\begin{cases} 0.5-0.5\cos\left(\frac{2\pi}{L}n\right) & 0\leq n\leq L-1 \\ 0 & \text{otherwise} \end{cases}, \quad (3)$$

where L is the length or the number of samples of the Hanning time window. The data that are stored in the Hanning time window are evaluated for spectral computation, which involves computing the discrete Fourier transform (DFT) using the FFT algorithm, as described in the next section.

3. NOISE REMOVAL IN FREQUENCY DOMAIN AND CONVERSION TO THE TIME DOMAIN USING IFFT

In this section, we present our algorithm for the spectral subtraction of noise from a speech signal [6] . A flowchart of this algorithm is shown in Figure 3. In the subsection below, we explain the role of the various blocks of this flowchart.

3.1. Speech and Non-Speech Activity Frame

Previously we discussed how we recorded two signals, a clean speech S[n] and a vacuum cleaner noise N[n]. The vacuum cleaner noise was digitally added to the clean speech to form a noisy speech X[n].

Appending X[n] to N[n], we composed the signal NX[n] shown in Figure 4(a). The first part of the signal is composed of the non-speech activity that contains the stationary noise of the vacuum cleaner with 28,446 samples. The second part of the signal is composed of the speech activity that contains the noisy speech "Real graph" with 28,446 samples. Both speech and non-speech activity spectrums were computed using the FFT.

The spectrum of the speech activity containing the noisy speech time frame is denoted as

$$\tilde{X}[k]=S[k]+N[k], \quad (4)$$

where S[k] is the spectrum of the clean speech "Real graph" and N[k] is the spectrum of the vacuum cleaner noise. The spectrum NX[k] of the signal composed of the vacuum noise followed by the noisy speech is shown in Figure 4(b).

3.2. Computing the Average Magnitude of the Noise Spectrum during Non-Speech Activity

The non-speech activity contains 221 time frames with 256 frequency values. After computing the non-speech activity spectrum, we calculate the average of the noise magnitude spectrum for each frequency.

$$\mu(k)=E\{|N[k]|\}, \quad (5)$$

where E is the average value operator. In the next subsection, we explain the role of μ (k).

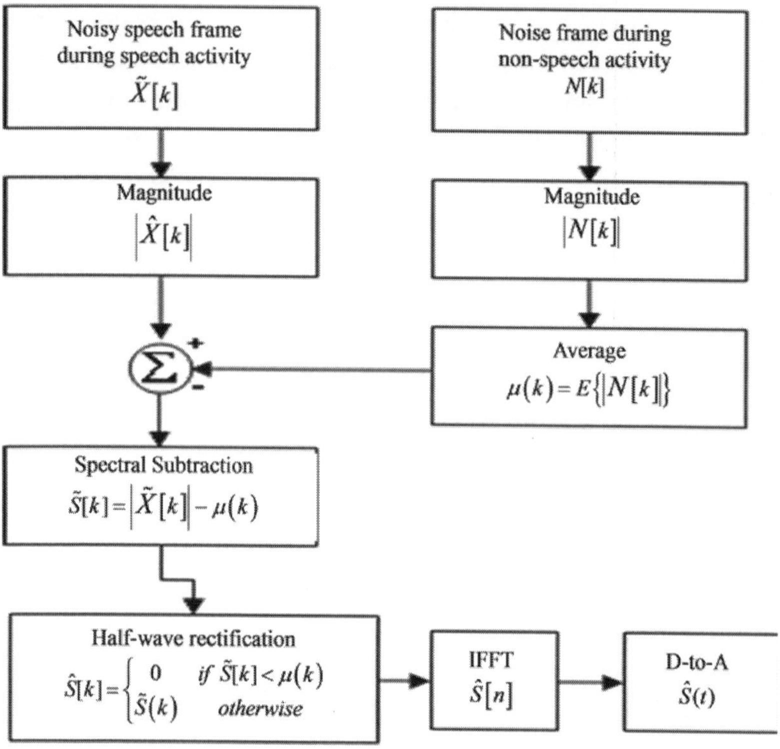

Figure 3. Spectral subtraction noise removal flowchart.

3.3. Noise Removal by Subtracting Average Magnitude of Noise Spectrum

The speech activity which is the noisy speech "Real graph" contains 221 rows of time frames and 256 columns of frequency values. The average magnitude of the noise spectrum is subtracted from the noisy speech spectrum resulting in the signal

$$\tilde{S}[k]=|\tilde{X}[k]|-\mu(k) \quad (6)$$

Figure 5 shows the noisy speech frame after subtracting the average noise magnitude spectrum.

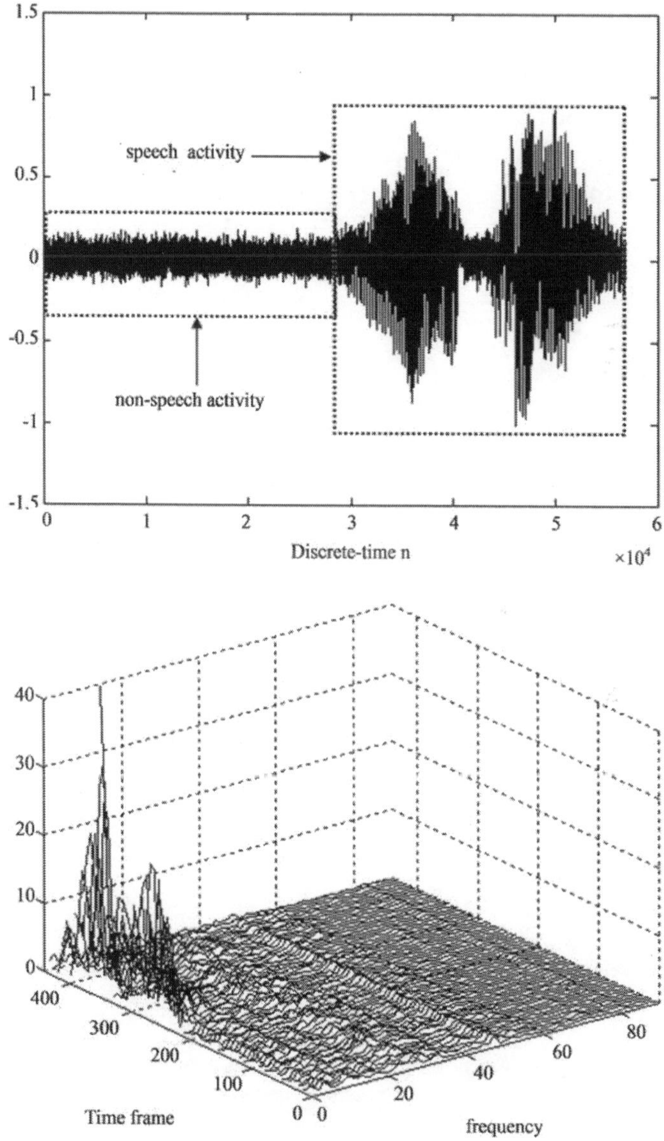

Figure 4. (a) The signal NX[n] composed of N[n] followed by X[n]; (b) Spectrum NX[k] of NX[n].

3.4. Half-Wave Rectification

In some cases, for each frequency ω, the value of the average magnitude of the noise spectrum is larger than the magnitude of the noisy speech spectrum. This results in negative values after subtracting the average magnitude of the noise spectrum from the noisy speech spectrum. Half-wave rectification consists in replacing those negative values with zero resulting in the signal

$$\hat{S}[k]=\begin{cases}0 & \text{if}\,\tilde{S}[k]<\mu(k)\\ \tilde{S}[k] & \text{otherwise}\end{cases} \quad . \tag{7}$$

Figure 6 shows the spectrum of the speech frames after subtracting the average noise magnitude and halfwave rectification.

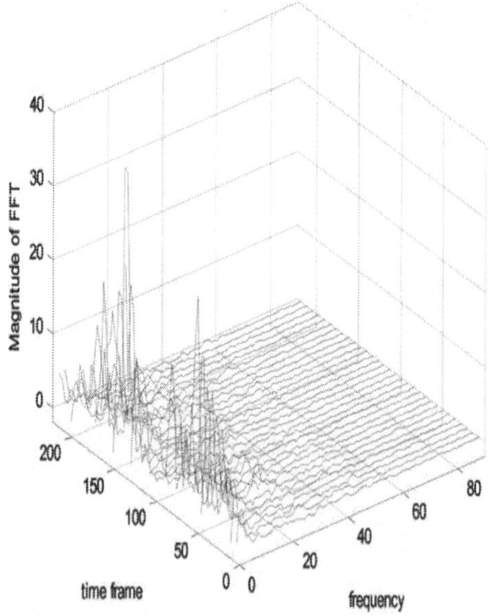

Figure 5. Spectrum of the noisy speech frames after subtracting the average noise magnitude spectrum.

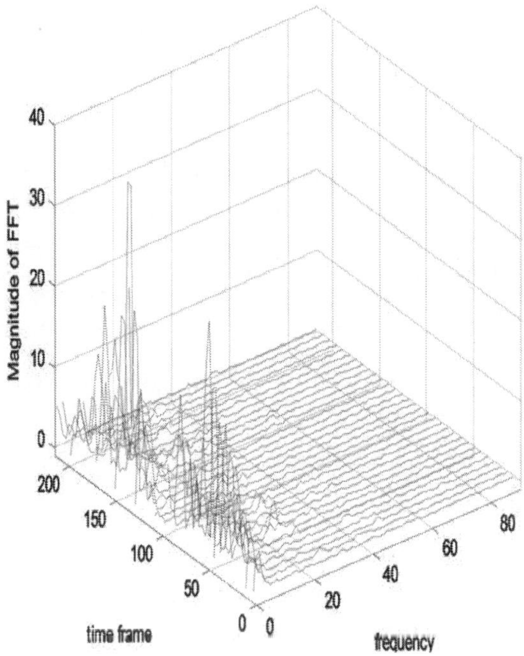

Figure 6. Spectrum of the speech frames after subtracting average noise magnitude and half-wave rectification.

3.5. Reconstruction of the Noisy Speech "Real Graph" Using Inverse Fast Fourier Transform

The flowchart blocks detailed above complete the noise removal algorithm. The conversion from frequency domain to discrete-time domain using the IFFT [7] of the signal $\tilde{S}[k]$ results in the signal $\tilde{S}[n]$ calculated as follows:

$$\hat{S}[n] = \frac{1}{N}\sum_{k=1}^{N}\hat{S}[k]\,e^{j(2\pi K/N)n} \qquad 0 \le n \le N-1 \tag{8}$$

3.6. Transformation of Noisy Speech "Real Graph" in Real Time Using D-to-A Converter

After converting the reconstructed speech signal $\hat{S}[n]$ from the frequency domain to the discrete time domain using IFFT, the Digital-to-Analog converter transforms $\hat{S}[n]$ back to the real-time speech signal $\hat{S}(t)$. We implemented this algorithm using Matlab. The sound corresponding to $\hat{S}(t)$ was generated also using Matlab. Successful results were obtained. Noise was effectively reduced from the noisy speech "Real graph", which confirms the validity of our noise-removal algorithm.

4. IMPROVING THE SPECTRAL SUBTRACTION NOISE REMOVAL DESIGN

4.1. Frames Averaging

In this section, frames averaging technique was applied in order to improve the performance of the spectral subtraction noise removal design [9] . Frames averaging are represented as an optional step between computing the average magnitude of the noise spectrum and subtracting this average from the magnitude of the noisy speech frames. When applied, frames averaging involve using the magnitude average of several frames of the noisy speech rather than one frame at a time. We limited this research to averaging either three or six consecutive frames. Bigger numbers could result in decreasing in speech intelligibility [6] .

4.2. Varying the Lengths of the Half-Overlapped Data Buffers and Hanning Time Window

In this subsection we investigate the effect that varying the lengths of the half-overlapping data buffer and Hanning time window have on improving the noise removal design. The overlapping of the data buffers was varied between one-half and one-fourth overlapping and the Hanning time window length was varied from 256 points to 128 points and 512 points.

4.3. Statistical Error Analysis

The original algorithm design consisted of no frame averaging with half-overlapped data buffers and 256 points Hanning time windows. Combining those techniques lead to 17 different design configurations shown in Table 1, Table 2, Table 3.

In order to evaluate the improvement in noise removal, we used the Speech to Noise Ratio (SNR) [8] defined as

$$SNR_{dB} = 20\log\left(\frac{\hat{S}_{RMS}}{N_{RMS}}\right), \tag{9}$$

where $\hat{S}_{RMS}$ is the root-mean-square (RMS) of the reconstructed speech signal $\hat{S}[n]$ after noise removal calculated as follows

$$\hat{S}_{RMS} = \sqrt{\frac{1}{M}\sum_{n=1}^{M}\hat{S}^2[n]}, \tag{10}$$

and N_{RMS} is the RMS of the vacuum noise N[n] calculated as

$$N_{RMS} = \sqrt{\frac{1}{M}\sum_{n=1}^{M} N^2[n]} \quad (11)$$

Table 1. Noise removal design using 256 points Hanning window.

Various noise removal design configurations	*SNR* of the modified reconstruction signal (dB)	Δ_{SNR} (dB)
256 points Hanning window One-half overlapped data buffers 3 frames averaging	12.4914	−0.5334
256 points Hanning window One-half overlapped data buffers 6 frames averaging	12.4538	−0.5710
256 points Hanning window One-fourth overlapped data buffers No frame averaging	13.3619	0.3371
256 points Hanning window One-fourth overlapped data buffers 3 frames averaging	12.3457	−0.6791
256 points Hanning window One-fourth overlapped data buffers 6 frames averaging	12.6906	−0.3342

Table 2. Noise removal design using 128 points Hanning window.

Various noise removal design configurations	*SNR* of the modified reconstruction signal (dB)	Δ_{SNR} (dB)
128 points Hanning window One-half overlapped data buffers No frame averaging	12.9805	−0.0443
128 points Hanning window One-half overlapped data buffers 3 frames averaging	12.2872	−0.7376
128 points Hanning window One-half overlapped data buffers 6 frames averaging	12.2080	−0.8168
128 points Hanning window One-fourth overlapped data buffers No frame averaging	13.2412	0.2164
128 points Hanning window One-fourth overlapped data buffers 3 frames averaging	10.5704	−2.4544
128 points Hanning window One-fourth overlapped data buffers 6 frames averaging	12.2584	−0.7664

We considered as reference SNR (SNR_{ref}) the ratio of the RMS of the reconstructed speech signal "Real graph" in Section 3 to the RMS of the vacuum noise. Subsequently, the improvement in noise removal Δ_{SNR} was evaluated by subtracting the SNR_{ref} from the SNR of each of the 15 different design configurations in Table 1, Table 2, Table 3 as follows

$$\Delta_{SNR} = SNR - SNR_{ref} \quad (12)$$

Improvement indeed occurs whenever Δ_{SNR} is positive. In this application, SNR_{ref} was equal to 13.0248 dB, which lead to the Δ_{SNR} values shown in Table 1, Table 2, Table 3.

Table 3. Noise removal design using 512 points Hanning window.

Various noise removal design configurations	*SNR* of the modified reconstruction signal (dB)	Δ_{SNR} (dB)
512 points Hanning window One-half overlapped data buffers No frame averaging	13.1032	0.0784
512 points Hanning window One-half overlapped data buffers 3 frames averaging	13.0085	−0.0163
512 points Hanning window One-half overlapped data buffers 6 frames averaging	12.9394	−0.0854
512 points Hanning window One-fourth overlapped data buffers No frame averaging	5.1722	−7.8526
512 points Hanning window One-fourth overlapped data buffers 3 frames averaging	5.8646	−7.1602
512 points Hanning window One-fourth overlapped data buffers 6 frames averaging	6.5488	−6.476

5. CONCLUSIONS

In this research, noise removal from noisy speeches has been studied and analyzed. The study includes methods for removing noise from noisy speeches using spectral subtraction.

Real-time data were sampled using a digital sound recorder system that converted both clean speech "Real graph" and vacuum cleaner noise from analog signals to digital signals at a sampling rate of 44.1 kHz. Noisy speech was digitally generated by corrupting the data of the clean speech "Real graph" with the data of the vacuum cleaner noise. Noise removal in the time domain was not successful. However, in the frequency domain, noise was successfully removed from the noisy speech. Prior to removing noise in the frequency domain, the spectrums of speech and non-speech activities were computed using the FFT of Hanning time-windowed data buffers. Removing noise requires an approximation of the noise during speech activity. The approximation of the noise was obtained by taking the average magnitude of the noise spectrum during non-speech activity. The average magnitude of the noise spectrum during non-speech activity was subtracted from the noisy speech spectrum during speech activity.

Our initial noise removal design consisted of no frame averaging with half-overlapped data buffers and 256 points Hanning time windows. The corresponding reference signal to noise ratio was equal to 13.0248 dB. We then tested a total of 17 modified noise-removal designs in search for the best configuration. Results showed that using any combination of half-overlapped and one-fourth-overlapped data buffers with 128, 256, and 512 points, Hanning windows and three frames or six frames averaging did not improve the performance of the denoising algorithm. However, using one-fourth-overlapped data buffers with 256 points Hanning windows and no frames averaging resulted in the greatest improvement differential SNR in the amount of 0.3371 dB, leading to most noise removal from the noisy speech "Real graph". Thus we consider that our goal of denoising noisy speech signals has been successfully achieved.

REFERENCES

1. Hymavathy, K.P. and Janardhanan, P. (2013) Noise Filtering in Speech Using Frequency Response Masking Technique. International Journal of Emerging Trends in Engineering and Development, 2.

2. Muangjaroen, S. and Yingthawornsuk, T. (2012) A Study of Noise Reduction in Speech Signal Using FIR Filtering. Proceedings of the International Conference on Advances in Electrical and Electronics Engineering, Pattaya, 13-15 April 2012.

3. Kumar, T.L. and Rajan, K.S. (2012) Noise Suppression in Speech Signals Using Adaptive Algorithms. International Journal of Engineering Research and Applications, 2, 718-721.

4. Aggarwal, R., Singh, J.K., Gupta, V.K., Rathore, S., Tiwari, M. and Khare, A. (2011) Noise Reduction of Speech Signal Using Wavelet Transform with Modified Universal Threshold. International Journal of Computer Applications, 20, 15-19.

5. Verteletskaya, E. and Simak, B. (2010) Speech Distortion Minimized Noise Reduction Algorithm. Proceedings of the World Congress on Engineering and Computer Science, Vol. I, San Francisco, 20-22 October 2010.

6. Boll, S.F. (1979) Suppression of Acoustic Noise in Speech Using Spectral Subtraction. IEEE Transactions on Acoustic, Speech and Signal Processing, 27, 113-120.

7. Rabiner, L.R. and Schafer, R.W. (1978) Digital Processing of Speech Signals. Prentice Hall, Upper Saddle River.

8. Quantieri, T.F. (2001) Discrete-Time Speech Signal Processing: Principles and Practice. Prentice Hall, Upper Saddle River.

9. Allen, J. (1977) Short Term Spectral Analysis, Synthesis, and Modification by Discrete Fourier Transform. IEEE Transactions on Acoustic, Speech and Signal Processing, 25, 235-238.

CHAPTER 6

Denoising of Medical Images Using Multiwavelet Transforms and Various Thresholding Techniques

Abdullah Al Jumah[1], Mohammed Gulam Ahamad[1], Syed Amjad Ali[2]

[1]Department of Computer Engineering, College of Computer Engineering & Science, Salman bin Abdulaziz University, Al-Kharj, Saudi Arabia;

[2]Department of Electronics & Communication Engineering, Lords Institute of Engineering & Technology, Hyderabad, India.

ABSTRACT

The problem of estimating an image corrupted by additive white Gaussian noise has been of interest for practical reasons. Non-linear denoising methods based on wavelets, have become popular but Multiwavelets outperform wavelets in image denoising. Multiwavelets are wavelets with several scaling and wavelet functions, offer simultaneously Orthogonality, Symmetry, Short support and Vanishing moments, which is not possible with ordinary (scalar) wavelets. These properties make Multiwavelets promising for image processing applications, such as image denoising. The aim of this paper is to apply various non-linear thresholding techniques such as hard, soft, universal, modified universal, fixed and multivariate thresholding in Multiwavelet transform domain such as Discrete Multiwavelet Transform, Symmetric Asymmetric (SA4), Chui Lian (CL), and Bi-Hermite (Bih52S) for different Multiwavelets at different levels, to denoise an image and determine the best one out of it. The performance of denoising algorithms and various thresholding are measured using quantitative performance measures such as, Mean Square Error (MSE), and Root Mean Square Error (RMSE), Signal-to-Noise Ratio (SNR), Peak Signal-to-Noise Ratio (PSNR). It is found that CL Multiwavelet transform in combination with modified universal thresholding has given best results

KEYWORDS

Multiwavelets; Noise; Thresholding; Additive White Gaussian Noise; Signal-to-Noise Ratio; Discrete Multiwavelet Transforms; Chui Lian; Symmetric Asymmetric Multiwavelet Transform; Bi-Hermite Multiwavelet Transform; Modified Universal Thresholding

1. INTRODUCTION

Digital images play an important role both in daily life applications such as satellite television, Magnetic Resonance Imaging (MRI), computer tomography as well as in areas of research and technology such as geographical information systems and astronomy. Data sets collected by image sensors are generally contaminated by noise. Imperfect instruments, problems with the data acquisition process, and interfering natural phenomena can all degrade the data of interest. Furthermore, noise can be introduced by transmission errors and compression. Thus, denoising is often a necessary and the first step to be taken before the image data is analysed. It is necessary to apply an efficient denoising technique to compensate for such data corruption. Image denoising still remains a challenge for researchers because noise removal introduces artifacts and causes blurring of the images. This paper describes different methodologies for noise reduction giving an insight as to which algorithm should be used to find the most reliable estimate of the original image data, given its degraded version. Noise modeling in images is greatly affected by capturing instruments, data transmission media, image quantization and discrete sources of radiation. Different algorithms are used depending on the noise model. Most of the natural images are assumed to have additive random noise which is modeled as a Gaussian.

The developments in wavelet theory have given rise to the wavelet thresholding method, for extracting an image from noisy data. Multiwavelets, wavelets with several scaling functions, have recently been introduced and they offer simultaneous orthogonality, symmetry and short support, which is not possible with ordinary wavelets, also called scalar wavelets. This property makes Multiwavelets more suitable for various image processing applications, especially denoising.

Donoho and Johnstone pioneered the theoretical formalization of filtering additive white Gaussian noise (of zero mean and standard deviation) via thresholding the decomposed coefficients. A decomposed coefficient is subjected to a given threshold and is set to zero if its magnitude is less than the threshold; otherwise, it is kept or modified, depending upon the thresholding rule. The choice of threshold determines to a great extent the efficiency of denoising algorithm.

Denoising procedure is depicted in Figure 1.

Step 1: Original image is a medical tomographic image (head phantom) to which noise components like additive white Gaussian noise is added to create a noisy image → | → |.

Step 2: Pre-filtering is done on the noisy image to convert the scalar coefficients of an image into vector coefficients for Multiwavelet decomposition.

Step 3: It deals with decomposition of pre-filtered image using various Multiwavelets like GHM, CL, SA4, BiHermite52S, and their respective Multiwavelet transforms.

Step 4: Thresholding methods are used to remove noise from decomposed image. Here we apply Multiwavelet thresholds such as soft, hard, universal, modified, fixed and multivariate thresholds.

Step 5: By applying inverse Multiwavelet transforms (IMWT), to thresholded coefficients, we get the denoised vector output image.

Step 6: Post filtering is done on reconstructed image to get back the denoised image coefficients in scalar form.

2. Theoretical Aspects of Multiwavelets

The idea of Multiwavelet originates from the generalization of scalar wavelets [1,2]. Instead of one scaling and one wavelet function, multiple scaling and multiple wavelet functions are used. This leads to more degree of freedom in constructing Multiwavelets. Therefore, opposed to scalar wavelets, properties such as orthogonality, symmetry, higher order of vanishing moments, compact support can be gathered simultaneously in Multiwavelets. Multiwavelets are of mainly two types: 1) Orthogonal type such as Geronimo-Hardin-Massopust (GHM),

Symmetric Asymmetric (SA4), Chui-Lian (CL); and 2) BiOrthogonal type such as Bi-Orthogonal Hermite (Bih52S).

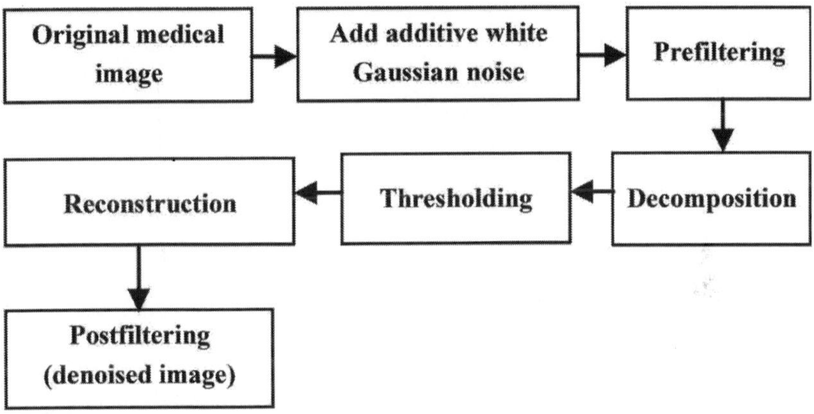

Figure 1. Block diagram of denoising using Multiwavelet transformation.

[3,4]. The scaling functions f_1 and f_2 are symmetric (linear phase) and they have short support (two intervals or less). The coefficients of Multiwavelets are 2 × 2 matrix. It retains the orthogonality of the Multiwavelets. The incoming signal is scalar type and is converted to vector type by using prefilter. The vector image is applied in discrete time to discrete Multiwavelet transform for low pass filtering, using low pass filter coefficients and down sampled (decimated) by 2, to get c_k coefficients and high pass filter coefficients are used for high pass filtering and down sampled by two to get d_k coefficients. This is the two band analysis bank. A perfect reconstruction synthesis bank recovers the image from the two down sampled outputs as shown in Figure 2 [5-7].

The sub-bands of a single level Multiwavelet decomposition is shown in Figure 3. It has 16 sub-bands of an image [8,9].

Multiwavelets are characterized with several scaling functions and associated wavelet functions as given in Equations (1) and (2) respectively [10-12].

$$\phi(t) = \Sigma c_k (2t - k) \tag{1}$$

$$W(t) = \Sigma d_k (2t - k) \tag{2}$$

where f(t) is a multiscaling function, W(t) is a Multiwavelet function.

A multi filter has two or more low pass filters. The purpose of this multiplicity is to achieve the following properties [1].

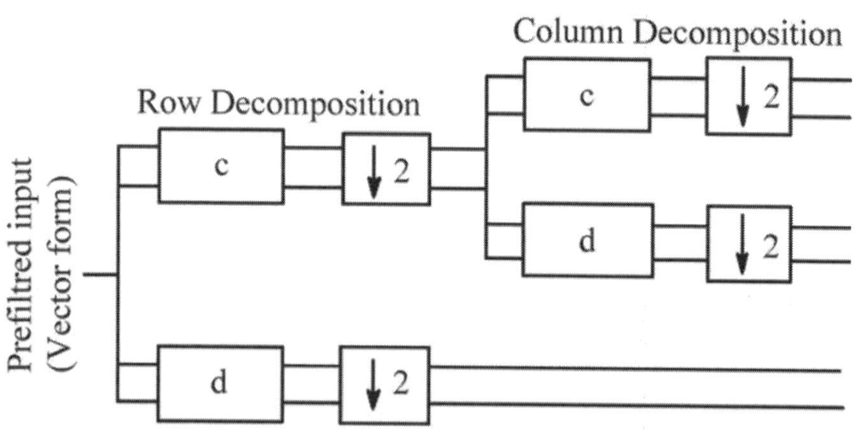

Figure 2. A multifilter bank with low pass filter iterated once.

L_1L_1	L_1L_2	L_1H_1	L_1H_2
L_2L_1	L_2L_2	L_2H_1	L_2H_2
H_1L_1	H_1L_2	H_1H_1	H_1H_2
H_2L_1	H_2L_2	H_2H_1	H_2H_2

Figure 3. Image sub-bands after one level of Multiwavelet decomposition.

Properties of $\boldsymbol{\Phi}_1$ and $\boldsymbol{\Phi}_2$: [13-15]

1) Symmetry: f_1 and f_2 have linear phase, i.e., they are even functions (after a shift of the origin).

2) Short Support: F_1 vanishes outside the interval [0, i]. Short support do not have much boundary problem. Long support must modify the function near boundaries. Wavelets have long support to achieve orthogonality. But Multiwavelets have short support and simultaneous orthogonality.

3) Second Order Accuracy: The scaling functions have second order accuracy, which establishes two vanishing moments.

4) Orthogonality: The translates $\phi_1(t-k) \& \phi_2(t-k)$ are all mutually orthogonal.

5) Matrix Multiscaling (Dilation) and Multiwavelet Equations: The coefficient "c_k" and "d_k" are 2 × 2 matrices multiplying vectors of scaling and wavelet functions as given in Equations (3) and (4) respectively.

$$\Phi(t)=\begin{bmatrix}\phi_1(t)\\ \phi_2(t)\end{bmatrix}=\sum_{k=0}^{3} c_k \begin{bmatrix}\phi_1(2t-k)\\ \phi_2(2t-k)\end{bmatrix} \tag{3}$$

$$\Psi(t)=\begin{bmatrix}\psi_1(t)\\ \psi_2(t)\end{bmatrix}=\sum_{k=0}^{3} c_k \begin{bmatrix}\psi_1(2t-k)\\ \psi_2(2t-k)\end{bmatrix} \tag{4}$$

3. THRESHOLDING TECHNIQUES

3.1. Hard Thresholding

It chooses coefficients greater than the given threshold λ and sets others to zero. It is unsuccessful in removing large noise coefficients.

The hard thresholding, operator is defined in Equation (5) as

$$F_h(x)=\begin{cases} x & \text{if } |x|\geq\lambda \\ 0 & \text{otherwise}\end{cases} \tag{5}$$

The transfer function of the same is shown in Figure 4.

If coefficient > λ, value attributes to original pixel, else its noise and we discard it.

3.2. Soft Thresholding

Soft thresholding yields smaller error than hard and is generally preferred over hard thresholding. Soft thresholding shrinks coefficients by the threshold λ towards zero. It is also called as shrinkage function. The soft thresholding operator is defined in Equation (6) as,

$$F_s(x)=\begin{cases} x-\lambda & \text{if } x\geq\lambda \\ 0 & \text{if } |x|<\lambda \\ x+\lambda & \text{if } x\leq-\lambda \end{cases} \tag{6}$$

The transfer function of the same is shown in Figure 5.

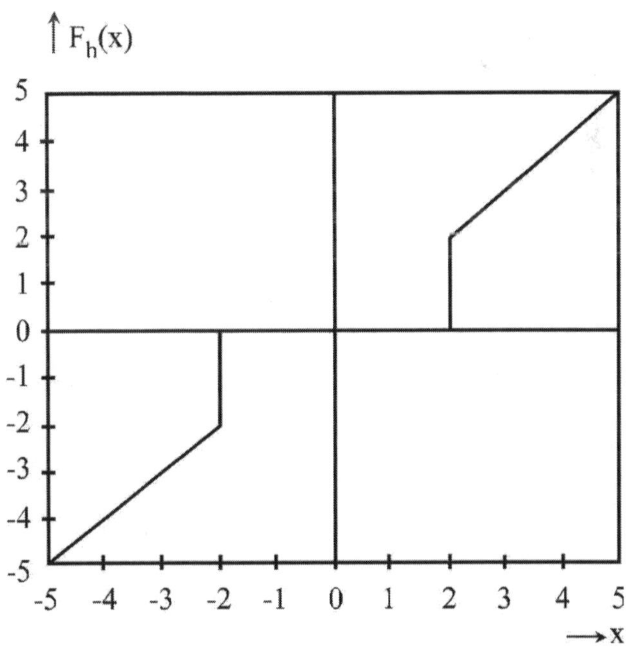

Figure 4. Hard thresholding.

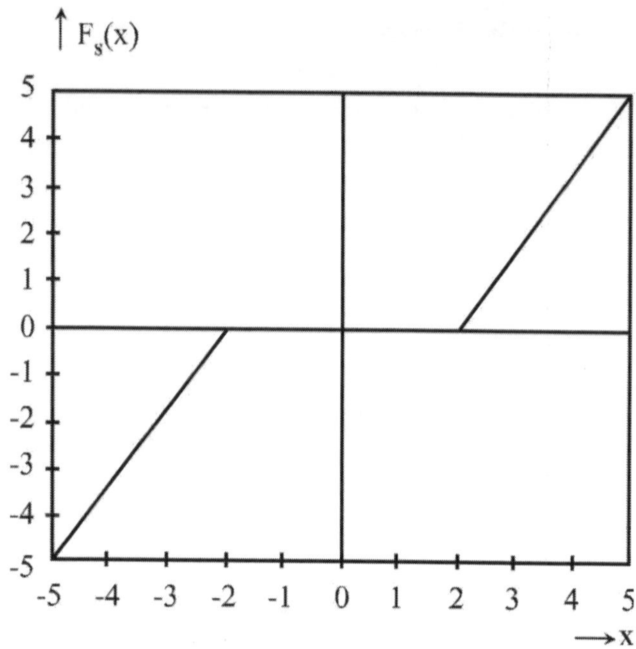

Figure 5. Soft thresholding.

3.3. Wavelet Thresholding [16,17]

Noise is generally present in the higher order frequency components obtained after wavelet and Multiwavelet decomposition. wavelet thresholding is the estimation technique which exploits the capabilities of wavelet transforms. Here small coefficients dominated by noise are set to zero, when they are below a certain threshold. In this way, noise can be removed.

3.3.1. Universal Threshold

The Universal threshold was introduced by Donoho and Johnston for scalar wavelets. It is given in Equation (7) as,

$$\lambda_{\text{universal}} = \sigma\sqrt{2\log N}, \qquad (7)$$

where σ is the standard deviation of the noise.

N is the sample size.

Same threshold is applied to all coefficients regardless of the decomposition level.

For images, N is chosen to be the number of pixels in a row rather than the total number of pixels in the image. It is observed that this choice for N results in higher SNR values.

3.3.2. Modified Universal Threshold

1) The modified universal threshold is given in Equation (8) as,

$$\lambda_{\text{mod}} = \sigma\sqrt{2\log 2N}, \qquad (8)$$

where σ is the standard deviation of the noise. N is the sample size.

2) Another modified universal threshold is given in Equation (9) as,

$$\lambda_N = \sigma\left(\sqrt{2\log N + 2\log\log N}\right) \qquad (9)$$

where N is the sample size.

3.3.3. New Thresholding

The new threshold is given in Equation (10) as,

$$\lambda_{\text{Newthr}} = \sigma\left(\sqrt{2}\log\log N\right) \tag{10}$$

3.3.4. Multivariate Thresholding

This method is based on parent-child relationship which exists between image pixels of various sub-bands of Multiwavelet decomposition. If the parent coefficient has a small value, then the children would likely have small values. If the parent coefficient has a large value, the child might also have large values.

3.4. Measuring Parameters

3.4.1. Mean Square Error (MSE)

The MSE between the original image I(x,y) and the reconstructed image $I'(x,y)$ is given in Equation (11) as,

$$\text{MSE} = \frac{1}{MN}\sum_{y=0}^{M-1}\sum_{x=0}^{N-1}\left[I(x,y) - I'(x,y)\right]^2 \tag{11}$$

where I(x,y) is the original image, $I'(x,y)$ is the approximated version, which is actually the decompressed image and $M*N$ represents the size of an image.

3.4.2. Signal-to-Noise Ratio (SNR)

It is another measure often used to compare the performance of reproduced images. It is given in Equation (12) as,

$$\mathrm{SNR} = 10\log_{10}\left(\frac{1}{MN}\sum_{y=0}^{M-1}\sum_{x=0}^{N-1}[I(x,y)]^2 \Big/ \mathrm{MSE}\right) \tag{12}$$

3.4.3. Peak Signal-to-Noise Ratio (PSNR)

It is more subjective qualitative measurement of distortion. For 8-bit image, it is given in Equation (13) as,

$$\mathrm{PSNR} = 10\times\log_{10}\left(\frac{(255)^2}{(\mathrm{MSE})}\right) \tag{13}$$

3.5. Experimental Results

The measuring parameters such as MSE, RMSE, PSNR and SNR values for various Multiwavelet transforms with repeated row prefilter and thresholding techniques are tabulated in Table 1.

3.5.1. Using Repeated Row Pre-Filtering Method

The measuring parameters such as MSE, RMSE, PSNR and SNR values for various Multiwavelet transforms with approximation prefilter and thresholding techniques are tabulated in Table 2.

3.5.2. Using Approximation Pre-Filtering Method

The measuring parameters such as MSE, RMSE, and PSNR and SNR values for GHM Multiwavelet transforms with various thresholding techniques are tabulated in Table 3.

The measuring parameters such as MSE, RMSE, and PSNR and SNR values for GHMAP Multiwavelet transforms with various thresholding techniques are tabulated in Table 4.

The measuring parameters such as MSE, RMSE, and PSNR and SNR values for CARDBAL2 Multiwavelet transforms with various thresholding techniques are tabulated in Table 5.

Denoising of tomographic image of phantom is shown in Figure 6.

Comparison of various Multiwavelet transformation with thresholding techniques with respect to parameters such as MSE, SNR, RMSE are depicted in the Figures 7-9 respectively.

Comparison of various measuring parameters before and after thresholding is depicted in Figure 10.

Comparison between various Multiwavelet transforms along with various thresholding techniques is given in Figure 11 and Comparison between various Multiwavelet transforms is given in Figure 12.

3.6. Conclusions

Multiwavelets became a focus of research partly because they made possible the construction of wavelet systems that are simultaneously orthogonal, symmetric and Finite Impulse Response. However, it has become clear that the implementation of the discrete Multiwavelet transform does not require prefilters. The performance of the Multiwavelets with different thresholding methods were investigated. We have as well modified a thresholding method to give better performance for the tomographic image used in our algorithms and named it as new threshold.

Table 1. Measurement of denoising parameters using various Multiwavelet transformation with repeated row prefilter and thresholding techniques.

Parameters	GHM		CL		SA4		BiH52S	
	BT	AT	BT	AT	BT	AT	BT	AT
New Threshold								
MSE	0.0025	0.0004	0.0025	0.0011	0.0025	0.0012	0.0025	0.0155
RMSE	0.0498	0.0269	0.0498	0.0335	0.05	0.0349	0.0502	0.1246
PSNR (dB)	74.1916	82.1668	74.1866	77.6343	74.1541	77.273	74.1102	66.2236
SNR (dB)	6.324	10.3117	6.3215	8.0453	6.3052	7.8647	6.2833	2.34
Universal Threshold								
MSE	0.0025	0.0007	0.0025	0.0003	0.0025	0.0008	0.0025	0.0166
RMSE	0.0499	0.0263	0.0501	0.0186	0.05	0.028	0.0499	0.1288
PSNR (dB)	74.11648	79.7172	74.1375	82.7634	74.1442	7912769	74.1866	65.9309
SNR (dB)	6.3106	9.0868	6.297	10.6101	6.3003	8.8167	6.3215	2.1936
Modified Universal Threshold								
MSE	0.0025	0.0006	0.0025	0.0003	0.0025	0.0005	0.0025	0.0166
RMSE	0.0503	0.0251	0.0499	0.0167	0.0502	0.023	0.05	0.129
PSNR (dB)	74.1604	80.1536	74.1706	83.6994	74.1248	80.8961	74.1467	65.9207
SNR (dB)	6.2816	9.3049	6.3135	11.0779	6.2908	9.6747	6.3016	2.1886
Multivariate Threshold								
MSE	0.0025	0.0008	0.0025	0.0004	0.0025	0.0009	0.0025	0.0166
RMSE	0.0499	0.0275	0.0498	0.0195	0.05	0.0298	0.0501	0.1289
PSNR (dB)	74.1616	79.3328	74.1816	82.2447	74.1556	78.6606	74.1863	65.9273
SNR (dB)	6.309	8.8947	6.319	10.3887	6.306	8.5585	6.297	2.1896
Fixed Threshold								
MSE	0.0025	0.0003	0.0025	0.0004	0.0025	0.0019	0.0025	0.0008
RMSE	0.0503	0.0186	0.0497	0.0257	0.0498	0.0439	0.0497	0.1278
PSNR (dB)	74.1036	82.7341	74.1957	79.9169	74.1816	75.2828	74.1957	79.2436
SNR (dB)	6.28	10.5956	6.3262	9.1871	6.319	6.8696	6.3262	8.8504

Table 2. Measurement of denoising parameters using various Multiwavelet transformation with approximation prefilter and thresholding techniques.

Parameters	GHM		CL		SA4		BIH52S	
	BT	AT	BT	AT	BT	AT	BT	AT
New Threshold								
MSE	0.0025	0.0008	0.0025	0.0025	0.0025	0.0005	0.0025	0.0038
RMSE	0.0499	0.0282	0.0499	0.0499	0.0502	0.0228	0.05	0.062
PSNR (dB)	74.1765	79.14	74.1644	74.1633	74.112	80.9821	74.156	72.2764
SNR (dB)	6.3165	8.7986	6.3104	6.3224	6.3165	8.7986	6.3062	5.3664
Universal Threshold								
MSE	0.0025	0.0038	0.0025	0.0023	0.0025	0.0021	0.0025	0.0131
RMSE	0.0499	0.0615	0.0499	0.0481	0.0499	0.0455	0.0501	0.1143
PSNR (dB)	74.1655	72.3592	74.203	74.4897	74.161	74.973	74.1257	66.967
SNR (dB)	6.3109	5.4078	6.3297	6.4731	6.3109	5.4078	6.2912	2.7117
Modified Universal Threshold								
MSE	0.0025	0.0034	0.0025	0.0025	0.0025	0.0014	0.0025	0.0087
RMSE	0.05	0.0581	0.0499	0.0499	0.0501	0.0378	0.0498	0.0933
PSNR (dB)	74.1518	72.8456	74.1467	74.1612	74.1351	76.5872	74.1943	68.7368
SNR (dB)	6.3109	5.4078	6.3083	6.3088	6.3109	5.4078	6.3254	2.9201
Multivariate Threshold								
MSE	0.0025	0.0043	0.0025	0.0023	0.0025	0.0023	0.0025	0.0135
RMSE	0.05	0.0653	0.05	0.0478	0.0502	0.048	0.0499	0.1163
PSNR (dB)	74.1437	71.8352	74.1937	74.5359	74.1199	74.5136	74.1766	66.8158
SNR (dB)	6.3001	5.1458	6.3016	6.4962	6.3001	5.1458	6.3166	2.3276
Fixed Threshold								
MSE	0.0025	0.0005	0.0025	0.0006	0.0025	0.0008	0.0025	0.0011
RMSE	0.0501	0.023	0.05	0.0251	0.05	0.028	0.0502	0.0329
PSNR (dB)	74.1417	80.895	74.1442	80.1489	74.1455	79.173	74.1201	77.7814
SNR (dB)	6.299	9.6758	6.3003	9.3027	6.299	9.6758	6.2883	8.119

Table 3. Measurement of denoising parameters using direct Multiwavelet transform (GHM) using various thresholding techniques.

Parameters	New Thr		Universal Thr		Modified Universal Thr		Multivariate Thr		Fixed Thr	
	BT	AT	BT	AT	BT	AT	BT	AT	BT	AT
MSE	0.0025	0.0011	0.0025	0.0036	0.0025	0.0025	0.0025	0.004	0.0025	0.001
RMSE	0.0502	0.0328	0.05	0.0596	0.0502	0.0503	0.0499	0.0633	0.0499	0.0321
PSNR (dB)	74.1246	77.8223	74.1529	72.6189	74.1131	74.094	74.165	69.506	74.1541	78.0126
SNR (dB)	6.2908	8.1397	6.3046	5.5377	6.2848	6.2752	6.3108	5.2826	6.3085	8.2347

Table 4. Measurement of denoising parameters using direct Multiwavelet transform (GHMAP) using various thresholding.

Parameters	New Thr		Universal Thr		Modified Universal Thr		Multivariate Thr		Fixed Thr	
	BT	AT	BT	AT	BT	AT	BT	AT	BT	AT
MSE	0.0025	0.0011	0.0025	0.0029	0.0025	0.0027	0.0025	0.0032	0.0025	0.001
RMSE	0.05	0.0325	0.0498	0.0538	0.05	0.0517	0.0501	0.0564	0.05	0.0321
PSNR (dB)	74.1855	77.5138	74.1406	73.4042	74.1584	73.8591	74.1346	73.1048	74.1467	78.0114
SNR (dB)	6.3109	8.1434	6.2985	5.9305	6.3075	6.1581	6.2955	5.7807	6.3015	8.2341

Table 5. Measurement of denoising parameters using Multiwavelet transform (CARDBAL2) with identity prefilter using various thresholding.

Parameters	New Thr		Universal Thr		Modified Universal Thr		Multivariate Thr		Fixed Thr	
	BT	AT	BT	AT	BT	AT	BT	AT	BT	AT
MSE	0.0025	0.0009	0.0025	0.0008	0.0025	0.0008	0.0025	0.0008	0.0025	0.001
RMSE	0.0498	0.0292	0.0499	0.0282	0.0497	0.0281	0.0501	0.0283	0.05	0.0312
PSNR (dB)	74.1816	78.8104	74.1675	79.1367	74.1957	79.1642	74.1352	79.0973	74.1504	78.2357
SNR (dB)	6.319	8.6334	6.3121	8.797	6.3262	8.8107	6.2958	8.7768	6.3034	8.346

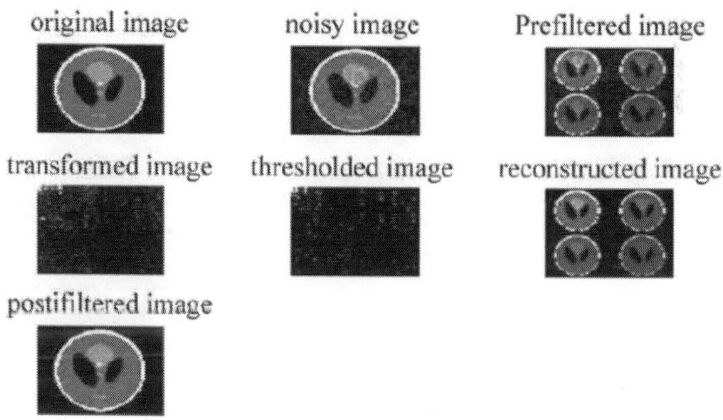

Figure 6. Denoising of a phantom using Multiwavelet transformation and thresholding techniques.

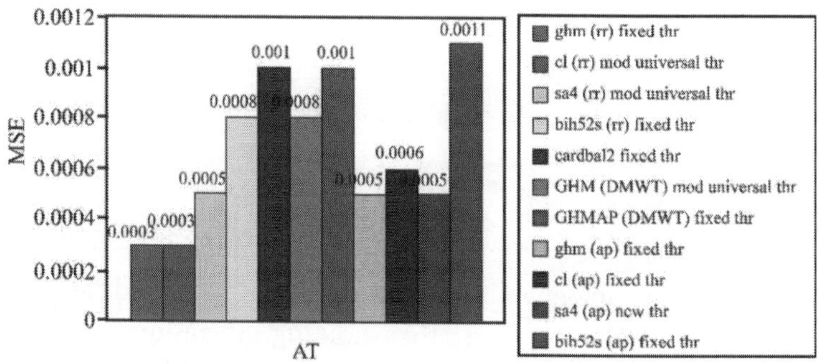

Figure 7. Comparison of various Multiwavelet transformation with thresholding techniques with respect to Mean Square Error.

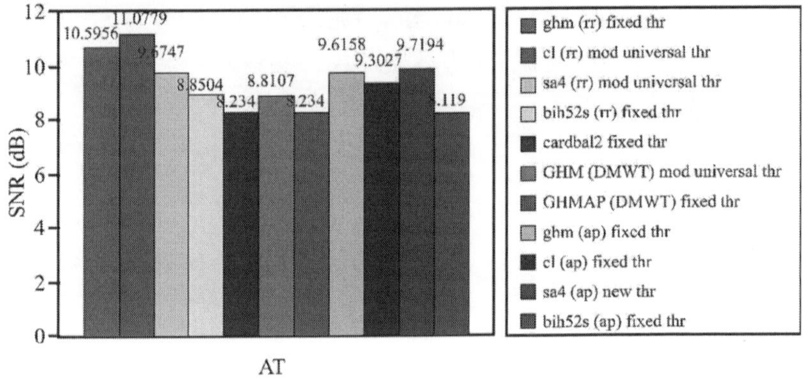

Figure 8. Comparison of various Multiwavelet transformation with thresholding techniques with respect to Signal-to-Noise Ratio.

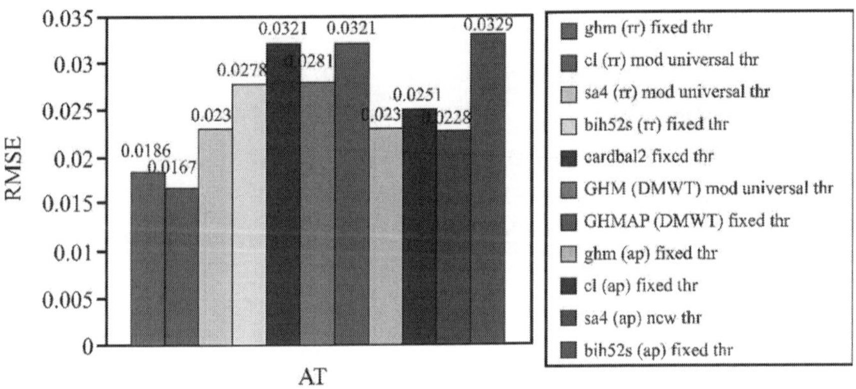

Figure 9. Comparison of various Multiwavelet transformation with thresholding techniques with respect to Root Mean Square Error.

To test the effectiveness of our algorithm, we added Gaussian noise to the 256 × 256 gray image phantom to, get noisy image and then applied Multiwavelet transforms and thresholding techniques to obtain denoised image. In this paper, different Multiwavelets like GHM, SA4, CL Bi-Hermite., and Multiwavelet transforms like Dec_2D, GHM and GHMAP with approximation and repeated row prefilters and various thresholding techniques like universal threshold, modified universal threshold, new threshold, multivariate threshold and fixed threshold is used to denoise the test image. Noise estimation parameters such as Root Mean Square Error (RMSE), Mean Square Error (MSE), Peak Signal to Noise Ratio (PSNR) and Signal to Noise Ratio (SNR) are used to evaluate the performance of algorithm.

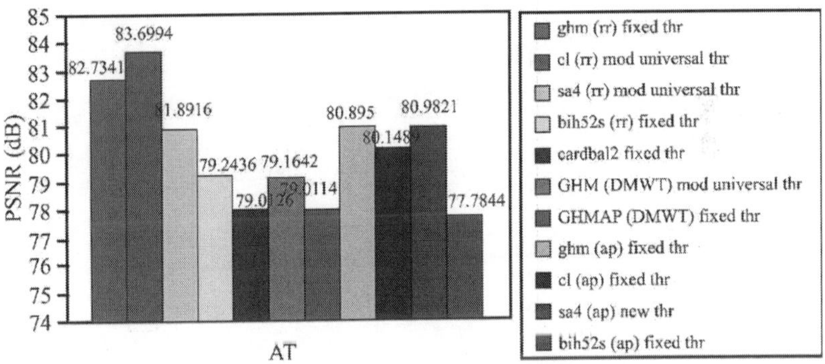

Figure 10. Comparison of various measuring parameters before and after thresholding.

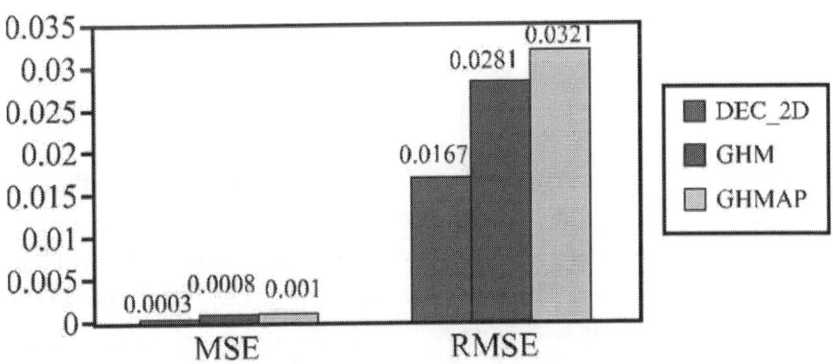

Figure 11. Comparison between various Multiwavelet transforms along with various thresholding techniques.

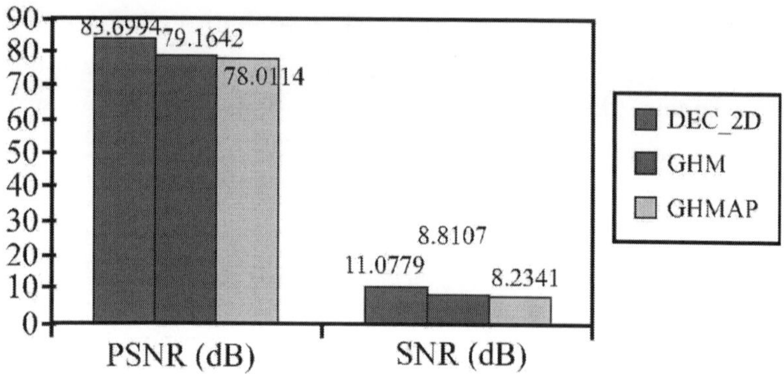

Figure 12. Comparison between various Multiwavelet transforms.

We have compared different Multiwavelets and thresholding techniques used. It is found that CL is the best Multiwavelet, when used with modified universal threshold and repeated row prefilter.

3.7. Future Scope

Nonetheless, there is always room for improvement. Since Multiwavelets are relatively a new subject of study, only a few construction methods for Multiwavelets are available. Most current filter available have two, three or fourth order of approximation. Future construction methods may add even higher order of approximation, while preserving the desirable features of current methods, would most likely result in multifilters that perform even better in image denoising and compression applications. Moreover the Multiwavelet systems available presently have the multiscaling and Multiwavelet coefficients which are 2 × 2 matrices. There is a possibility that in future many more Multiwavelet systems might be developed with matrix coefficients with higher order, which could provide even better results in the field of image denoising.

REFERENCES

1. G. Strang and V. Strela, "Short Wavelets and Matrix Dilation Equations," IEEE Transactions on Signal Processing, Vol. 43, No. 1, 1995, pp. 108-115.

2. G. Strang and V. Strela, "Orthogonal Multiwavelets with Vanishing Moments," Journal of Optical Engineering, Vol. 33, No. 7, 1994, pp. 2104-2107.

3. V. Strela, P. N. Heller, G. Strang, P. Topiwala and C. Heil, "The Application of Multiwavelet Filterbanks to Image Processing," IEEE Transactions on Image Processing, Vol. 8, No. 4, 1999, pp. 548-563.

4. V. Strela, and A. T. Walden, "Signal and Image Denoising via Wavelet Thresholding: Orthogonal and Biorthogonal, Scalar and Multiple Wavelet Transforms," Statistic Section Technical Report TR-98-01, Department of Mathematics, Imperial College, London, 1998.

5. B. R. Johnson, "Multiwavelet Moments and Projection Prefilters," IEEE Transactions on Signal Processing, Vol. 48, No. 11, 2000, pp. 3100-3108.

6. C. K. Chui and J. A. Lian, "A Study of Orthogonal Multiwavelets," Applied Numerical Mathematics, Vol. 20, No. 3, 1996, pp. 273-298.

7. D. P. Hardin and D. W. Roach, "Multiwavelet Prefilters-I: Orthogonal Prefilters Preserving Approximation Order P ≤ 2," IEEE Transactions on Circuits and Systems-II: Analog and Digital Signal Processing, Vol. 45, No. 8, 1998, pp. 1106-1112.

8. T. R. Downie and B. W. Silverman, "The Discrete Multiple Wavelet Transform and Thresholding Methods," IEEE Transactions on Signal Processing, Vol. 46, No. 9, 1998, pp. 2558-2561.

9. J. Y. Tham, L. X. Shen, S. L. Lee and H. H. Tan, "A General Approach for Analysis and Application of Discrete Multiwavelet Transforms,"

IEEE Transactions on Signal Processing, Vol. 48, No. 2, 2000, pp. 457-464.

10. K. Attakitmongcol, D. P. Hardin and D. M. Wilkes, "Multiwavelet Prefilters-Part II : Optimal Orthogonal Prefilters," IEEE Transactions of Image Processing, Vol. 10, No. 10, 2001, pp. 1476-1487.

11. G. Andria, F. Attivissimo, G. Cavone, M. Spadavecchia and T. Magli, "Denoising Filter to Improve the Quality of CT Images," Proceedings of IEEE Conference on Instrumentation and Measurement Technology, 5-7 May 2009, pp. 947-950.

12. L. X. Shen, H. H. Tan and J. Y. Tham, "Some Properties of Symmetric-Antisymmetric Orthonormal Multiwavelets," IEEE Transactions on Signal Processing, Vol. 48, No. 7, 2000, pp. 2161-2163.

13. V. Strela and A. T. Walden, "Orthogonal and Biorthogonal Multiwavelets for Signal Denoising and Image Compression," Proceedings of SPIE, Vol. 3391, 1998, pp. 96-107.

14. X. G. Xia, J. Geronimo, D. Hardin and B. Suter, "On Computations of Multiwavelets Transforms," Proceedings of SPIE, San Diego, July 1995.

15. X.-G. Xia, J. Geronimo, D. Hardin and B. Suter, "Design of Prefilters for Descrete Multiwavelets Transforms," IEEE Transactions on Signal Processing, Vol. 44, No. 1, 1996, pp. 25-35.

16. E. Bala and A. Ertuzun, "A Multivariate Thresholding Technique for Image Denoising Using Multiwavelets," EURASIP Journal on Applied Signal Processing, Hindawi Publishing Corporation, Cairo, 2005, pp. 1205-1211.

17. T.-C. Hsung, D. P.-K. Lun and K. C. Ho, "Optimizing the Multiwavelet Shrinkage Denoising," IEEE Transactions on Signal Processing, Vol. 53, No. 1, 2005, pp. 240-251.

CHAPTER 7

An Optical Tomography System Using a Digital Signal Processor

Ruzairi Abdul Rahim [1,*], Chiam Kok Thiam [1]
and Mohd Hafiz Fazalul Rahiman [2]

[1] Control & Instrumentation Engineering Department, Faculty of Electrical Engineering, Universiti Teknologi Malaysia, 81310 Skudai, Johor Bahru, Malaysia

[2] School of Mechatronic Engineering, Universiti Malaysia Perlis, 02600 Arau, Perlis, Malaysia

ABSTRACT

The use of a personal computer together with a Data Acquisition System (DAQ) as the processing tool in optical tomography systems has been the norm ever since the beginning of process tomography. However, advancements in silicon fabrication technology allow nowadays the fabrication of powerful Digital Signal Processors (DSP) at a reasonable cost. This allows this technology to be used in an optical tomography system since data acquisition and processing can be performed within the DSP. Thus, the dependency on a personal computer and a DAQ to sample and process the external signals can be reduced or even eliminated. The DSP system was customized to control the data acquisition process of 16×16 optical sensor array, arranged in parallel beam projection. The data collected was used to reconstruct the cross sectional image of the pipeline conveyor. For image display purposes, the reconstructed image was sent to a personal computer via serial communication. This allows the use of a laptop to display the tomogram image besides performing any other offline analysis.

KEYWORDS

Optical tomography; digital signal processor; image reconstruction

1. INTRODUCTION

Optical tomography involves the use of non-invasive optical sensors to obtain vital information in order to produce images of the dynamic internal characteristics of process system [4]. The use of optical sensors has the advantages of being conceptually straightforward and relatively inexpensive [7]. Its working principle involves projecting a beam of light through a medium from one boundary point and detecting the level of light received at another boundary point [9]. A typical optical tomography system usually uses a Data Acquisition System (DAQ) to perform data acquisition tasks and a host computer to carry out image reconstruction. However, only some of the functions provided by the DAQ were implemented. This is a waste of resources considering the high cost of a DAQ system. This paper discusses the implementation of DSP as the core processor in a parallel beam projection optical tomography system.

On the other hand, nowadays powerful Digital Signal Processors (DSP) running at Millions Instructions Per Second (MIPS) are available at very reasonable cost. A DSP, which is usually equipped with single cycle multiplication, is well suited for number crunching applications and optimized for digital signal processing algorithms. The tasks that were carried out by the DAQ and host computer such as data acquisition and image reconstruction can now be carried out by the DSP alone. However, the host computer is still needed for display purposes. Alternatively, a Liquid Crystal Display (LCD) could be utilized to display the results.

The full automation of solid-handling plant has frequently not been possible because of the lack of a basic flow meter [3]. The previously used technique to measure solid flow often involved removing the material for weighing which was complicated, expensive and did not allow continuous feedback for automatic control [3]. Many non intrusive techniques attempted to measure solids mass flow in pneumatic conveyor were elaborated Beck *et al.* [1].

Later, Xie *et al.* [13] made an effort to measure mass flow of solids in pneumatic conveyors by combining electrodynamics and capacitance transducers. Rahim *et al.* [10] investigated applications of fibre optics into optical tomography for flow concentration measurement. Chan [2] followed up and investigated mass flow rate measurement using optical tomography by relating the flow's concentration profile measurement with its mass flow rate. This method first performs calibration by recording the flow's concentration percentage plotted against mass flow rate measured using weight scale and stop watch. Based on their findings, the two variables had linear relation in light flow condition. However, this technique requires re-calibration each time a new flow material is used.

Cross correlation technique was proposed by Beck *et al.* [1] for flow velocity measurement. This function was successfully applied by Ibrahim and Green [5] for offline bubble flow velocity measurement using optical tomography. In 2003, Pang [7] implemented real time system for solid-gas flow concentration, flow velocity and mass flow rate measurement using optical tomography and data distribution system. The cross correlation techniques being used by these researches were based on time domain cross correlation function, which was very time consuming.

Pang [7] implemented four projections arrays (2 orthogonal and 2 rectilinear) in optical tomography system. The extra two layers (the Rectilinear layers) were used to filter the ambiguous image and smearing effects. However, analogue acquisition was used to acquire all the measurements. The obtained data from masking layer were used for comparison to predetermined threshold values. Using this method, valuable time was utilized for analogue to digital conversions. Furthermore, the number of sensors on the rectilinear was quite large. Although the research successfully implemented real time optical tomography system, it required four powerful personal computers and a network hub in order to implement data distribution system. The resulting system was bulky and not portable.

2. HARDWARE CONSTRUCTION

The topology is illustrated in Figure 1. The first module, which is the Signal Conditioning Unit, is responsible for the conditioning of incoming signals from the optical sensors. For Analogue Signal Conditioning, the incoming signals are from 64 optical sensors of the orthogonal layers. These signals are multiplexed into 16 ADC channels on the DSP system. On the other hand, 92 single bit binary streams from the Digital Signal Conditioning are sent to the embedded system using Serial Peripheral Interface (SPI). There are also signals from the DSP system to control the sampling process of these two modules.

The Main Controller Unit, as its name implies, acts as a main controller in processing the measurement data and producing required results. It is mainly used to perform the thousands of calculations required for image reconstruction algorithms, velocity profile construction and mass flow rate measurements. It also coordinates the communication between Signal Conditioning Unit and User Interface Unit.

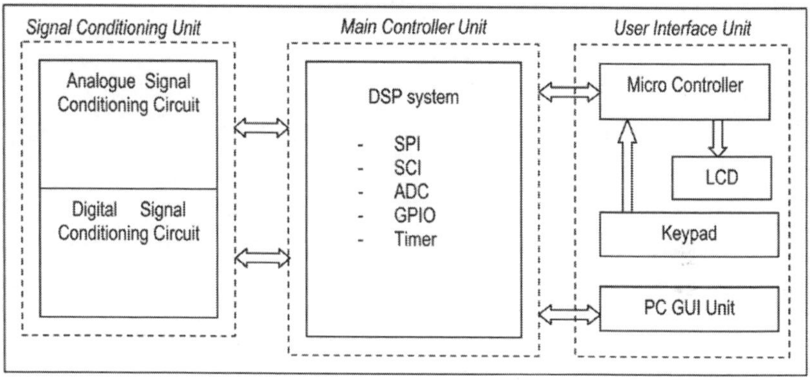

Figure 1. Block diagram of three main hardware modules.

2.1. Sensor Configuration

In this research, a fabricated optical sensor unit with sensors arranged in parallel beam projection has been implemented [8]. There are two layers of sensors being used, namely Upstream Layer and Downstream Layer to support the analysis of algorithm-based velocity profile measurement. At each layer, there are four arrays of sensors arranged in parallel beam projections. Two arrays of sensors, namely Orthogonal Sensors are arranged in such a way that their projected light beams are orthogonal (90°) to each other while another 2 arrays of sensors namely, Rectilinear Sensors are arranged so that their projected light beams are 45° to the orthogonal projections. This configuration is illustrated in Figure 2.

There are 16 sensors at each orthogonal projection (MU_0–MU_{15}, NU_0–NU_{15}, MD_0–MD_{15}, ND_0–ND_{15}) where altogether is a total of 64 sensors. The suffix U and D appended were to represent the Upstream and

Downstream layer. Another four arrays of 23 sensors (PU_0–PU_{22}, QU_0–QU_{22}, PD_0–PD_{22}, QD_0–QD_{22}) are arranged to provide rectilinear projections. This adds up 92 sensors for the rectilinear projections for Upstream and Downstream layers. In total, there are 156 sensors readings to be acquired at every single scan.

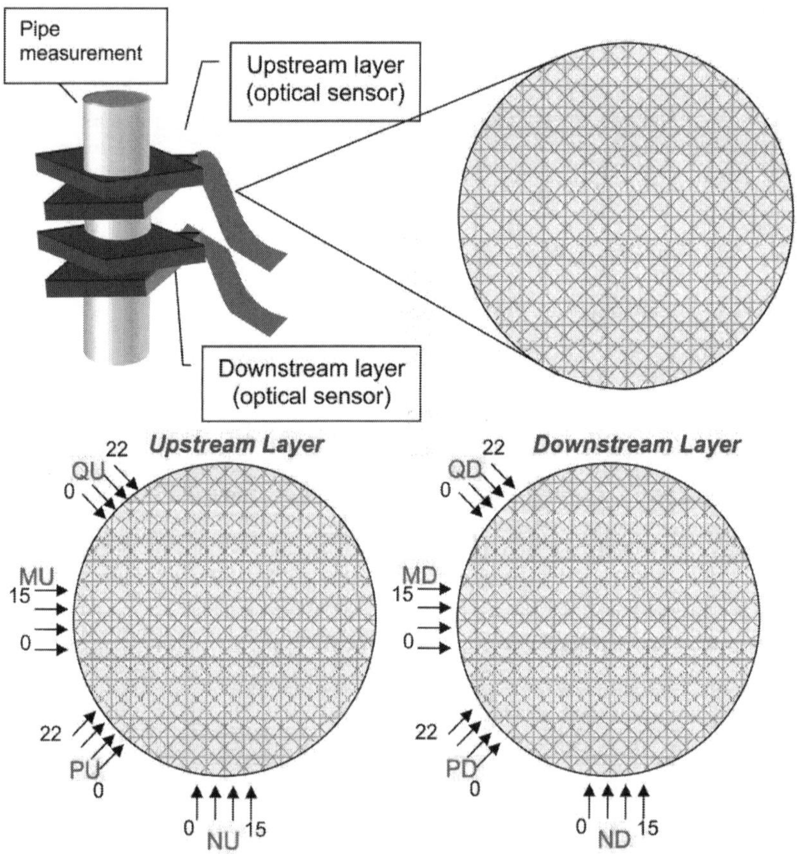

Figure 2. Sensor Unit configuration.

3. SIGNAL CONDITIONING UNIT

The Sensor Unit already includes LED driver circuits and amplifier circuits to convert the received light to electrical signals. Each sensor from the Sensor Unit outputs a current which is inversly proportional to the light intensity it measures. In other words, higher current is output when there is little light (object presence blocks the light beam) and lower current is output when there is more light (no object blocking the light beam). This way, software overhead is reduced during image reconstruction as the presence of object is proportional with the voltage measured. In work by Chan [2], optical attenuation model was used, where object presence resulted in voltage loss. The processing software then subtracted the loss from the expected full voltage to obtain the correct readings. This step was not required in this research.

All the signal amplifiers on the Sensor Unit are identical. Only resistors must be added to the outputs of the Sensor Unit to have a voltage drop for measurement. The circuit for a single sensor is shown in Figure 3. In this circuit, R2, R3 and R4 are used as simple current to voltage converter. R3 is a variable resistor, specifically used to adjust the scale of Vout. The value of 1MΩ is about 12.5% of the total drop voltage. This allows adjustment to calibrate optical sensors.

In designing the signal conditioning circuits, there are two considerations that need to be observed. The first consideration is the data acquisition rate. In total, there are 156 sensors readings to be acquired at every single scan. This obviously causes the measurement to take very long time, especially using a single digitizer with low conversion rate. Using a low conversion rate digitizer, an array of sample and hold circuit is required to acquire and hold the readings of a single scan [7]. This is because the conversions occur sequentially and by the time the digitizer converts the next sensor, the current signal is not from the same time as captured by the previous sensors. This may result in tilted or distorted image. Since there are 156 sensors on the sensor unit, using sample and hold circuits for each sensor will also take up a lot of board space.

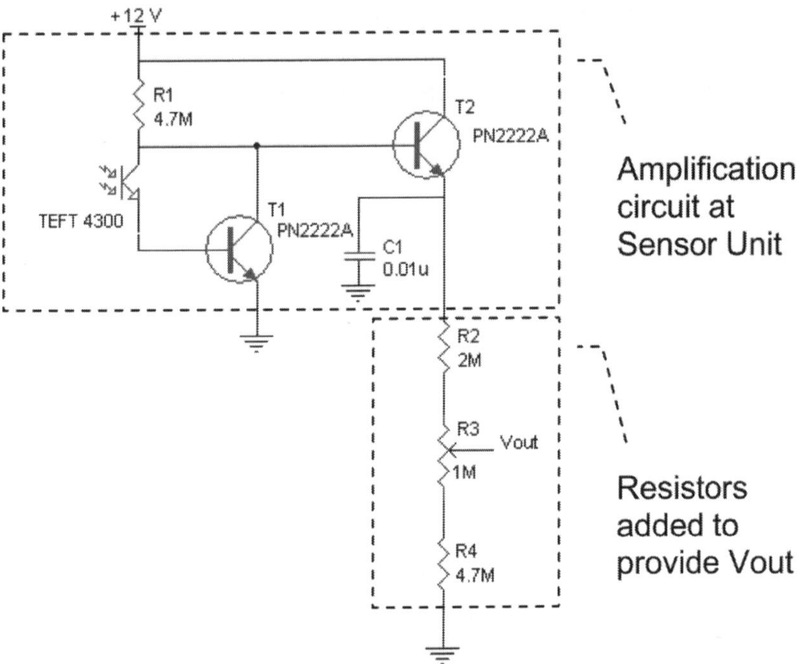

Figure 3. Amplification circuit for each sensor.

Another approach is to speed up measurement time by using dedicated low conversion rate digitizer per sensor or per group of sensors. The digitizers can be triggered to convert many signals simultaneously so that the need of sample and hold circuits can be eliminated. However, this approach is expensive and still occupies a lot of board space.

As a solution, a high conversion rate ADC was implemented in this research to overcome the said issues. The use of a single high conversion rate ADC required analogue multiplexers to multiplex respective signals for conversion. In this research, the ADC being used was built-in inside the DSP chip. Internally, it was multiplexed to receive analogue signals from 16 external channels. Since there are more than 16 signals for conversion, additional analogue multiplexers still need to be designed.

As for the second consideration, a faster method compared to ADC for sampling the sensors on rectilinear projections is desired. This is because these sensors only provide masking of ambiguous image in Hybrid Reconstruction algorithm. Previously, these signals were converted using ADC and compared to a threshold value set in software [7]. If the measured signal value was above the threshold value, a TRUE value was obtained and vice versa. Then, the signal value was discarded. Therefore, a single bit binary representation is sufficient and this can be implemented using hardware voltage comparator which can perform this same task at a higher rate. This approach reduces the number of signals for ADC conversion from 156 signals to only 64 signals.

3.1. Analogue Signal Conditioning Circuit

Basically, the Sensor Unit has provided the amplifications to convert the light intensity to current. Resistors were added to the output of the amplifier circuits to generate voltage drop for measurement. The fabricated circuits on the sensor unit were based on optical path width model [7] where the current outputs correlate linearly with the width of the object that obstructs the light beam.

As mentioned previously, there are 16 sensors per projection and there are two orthogonal projections per layer. Therefore, there are 32 signals for Upstream Layer and 32 signals for Downstream Layer, totalling up to 64 signals for measurement. In this research, a built in ADC in a DSP of the Main Controller Unit was utilized to perform the signal acquisitions.

The ADC is multiplexed into two groups, namely Group A and Group B. Each group consists of eight channels, thus expanding the total channel inputs of the ADC to be 16 channels. Due to the fact that there are 64 signals to convert, the signals need to be multiplexed. Four dual 4 to 1 analogue multiplexers (4052) were used per layer to multiplex these signals. Using this configuration, all 64 channels measurements can be accomplished in four acquisitions. The dual 4 to 1 analogue multiplexer was also chosen instead of the 8 to 1 analogue multiplexer since there were 16 ADC channels that were available. The data acquisition process can also be controlled easily since only two input bits (Sel MSB and Sel LSB) need to be controlled by using GPIO port. The overall configuration is shown in Figure 4.

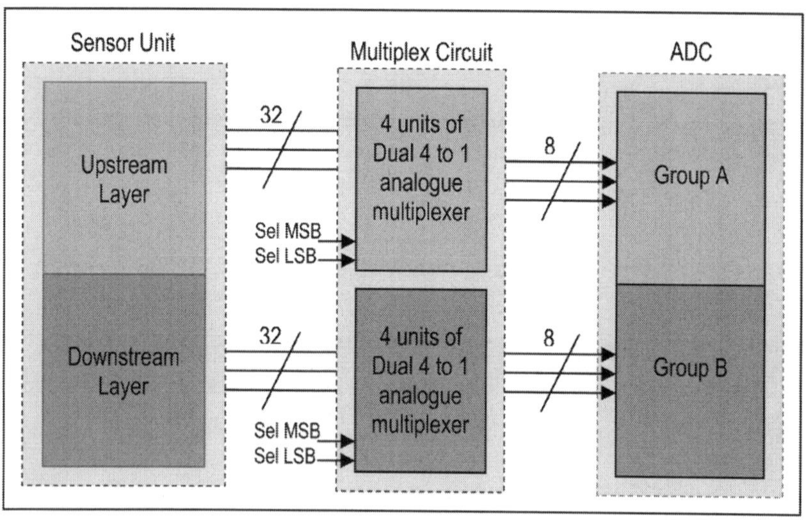

Figure 4. Interface between analogue signal conditioning and ADC.

3.2. Digital Signal Conditioning Circuit

As discussed in Section 3, improvement on the sampling method of the sensors for rectilinear projections was implemented using hardware voltage comparator. As the rectilinear projections were really used as a masking layer (to filter ambiguous image obtained from orthogonal projections), a single bit representation is sufficient (Either TRUE or FALSE). This is a trade off of more information (as in analogue values) with simpler and quicker processing. Each voltage comparator is implemented using TLC 2354 referenced to a voltage divider using variable resistor. These variable resistors were calibrated to a threshold value of about 25% of the steady state sensor output voltage (4V). This made sure that the representation was valid and not caused by false trigger such as noise and spikes. The following Equation 1 represents the value of the binary bit for different sensor voltage levels.

$$Vout = \begin{cases} LOW & ,Vin < 1\,V \\ HIGH & ,Vin \geq 1\,V \end{cases} \quad (1)$$

The circuit for each voltage comparator is shown in Figure 5. After the voltage comparisons were carried out, the binary bits need to be sent to the embedded system. This was conveniently achieved by converting them to serial bit stream and then reading the data using Serial Peripheral Interface (SPI). Each sensor output from the rectilinear projections was fed to a voltage comparator input (Vin).

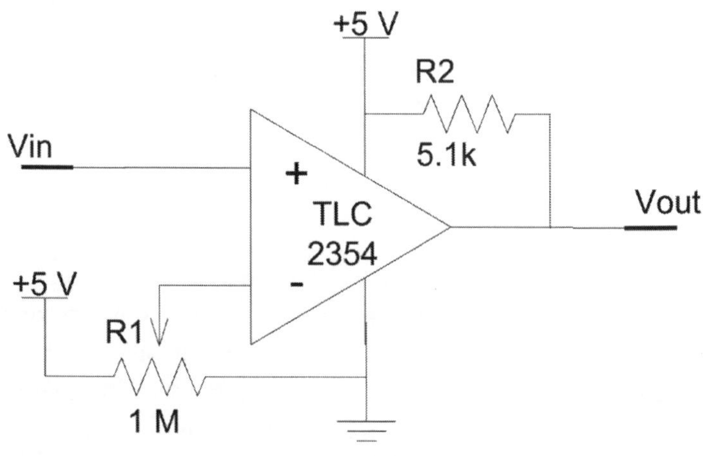

Figure 5. Voltage comparator circuit.

The resulting outputs (Vout) from the comparators were then connected to the data inputs of parallel to serial converter chips (74HC165) arranged in cascade. The embedded system only needed to load the parallel data bits by sending a pulse to the Parallel Load pin and generate clock signal to shift in the serial data. From the embedded system side of view, the Digital Signal Conditioning Circuit is merely a slave device with serial interface. This approach is less troublesome compared with using bus topology where buffers and more digital input and output lines are needed [12]. Besides that, the chip is able to support high speed data transfer. The overall topology of the digital signal conditioning circuit is shown in Figure 6.

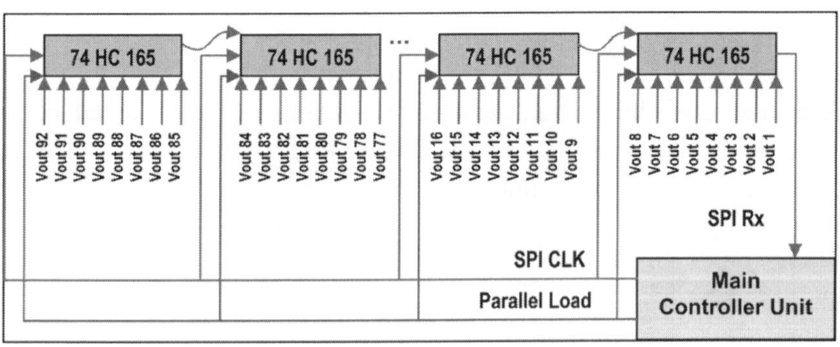

Figure 6. Digital Signal Conditioning topology.

4. MAIN CONTROLLER UNIT

The main controller unit is the brain of the whole system. The processing involved includes data acquisition, image reconstruction from measured sensors signals, flow analysis in terms of flow velocity and mass flow rate measurement and communication between modules. All these processes require intensive calculations and thus, the performance of this controller is very essential in affecting the overall system performance. Pang [8] developed an optical tomography that could perform image reconstruction in real time by using data distribution system where 3 sets of high performance PC (2.4 GHz, 512 MB RAM) were used. This research attempted to achieve comparable result using only single embedded system as its main controller unit.

Among the criteria that need to be fulfilled by the embedded system are high execution speed, low cost, support for standard communication protocols and memory expansion capability [8]. Micro processor based systems can offer high flexibility but requires more complicated design and other peripheral interfaces to support required functionalities. On the other hand, micro controller based systems offers many built in and ready to use standard peripherals but are usually not very flexible. Finally, a Digital Signal Processor based system provides relatively greater performance but lacks in terms of interfacing to the outside world.

A relatively new technology was introduced to the market lately and was known as the Hybrid Digital Signal Processor. The hybrid term was actually a combination of the micro controller and Digital Signal Processor technologies. As such, this architecture is actually a solution that benefits the best of both worlds. As this research requires both the peripherals offered by micro controller and high speed processing offered by Digital Signal Processor, the Hybrid Digital Signal Processor makes a very suitable choice as the platform for implementation in this research.

A comparison was carried out on some of the Digital Signal Processors offered by Texas Instruments Inc. [11] and the results are shown in Table 1. It can be concluded that the most suitable processor for this research is the one from the TMS320F28xx family. A device from this family, TMS320 F2812 fulfils almost all of the requirements needed. The unavailability of the Floating Point Unit is considered as a trade off for its low cost and can be justified by many of the supporting peripherals that come together in the chip.

Table 1. Comparison of available processors.

Family	Core Clock Speed (MHz)	Remarks
TMS320F28xx	150	Hybrid DSP, built in peripherals: UART (2), SPI, CAN 2.0, ADC (16.7 Msps), 128k Flash memory, GPIO, Dual 16x16 bit MAC
TMS320C55xx	300	Dual MAC, DMA, GPIO, Host Processor Interface (HPI), UART, ADC, I^2C,USB Full Speed (12 Mbps)
TMS320C64xx	600	Host Processor Interface (HPI), UART, GPIO (16 bits), 20 bit video port

5. DIGITAL SIGNAL PROCESSOR ARCHITECTURE

The chosen DSP (TMS320 F2812) as the main controller in the embedded system is a well equipped architecture with many essential peripherals built in and ready to be used. These peripherals include the internal high speed Analogue to Digital Converter (ADC), Serial Peripheral Interface (SPI), Serial Communication Interface (SCI) and 16 bit Timer.

5.1. User IO Unit

As mentioned previously, this unit allowed user to view Mass Flow Rate result and change certain settings of the Main Controller Unit. This unit consists of a microcontroller, LCD module and a 4×4 matrix keypad. By implementing this module, the control of LCD and keypad processing became transparent to the Main Controller Unit. Hence, the Main Controller Unit was able to communicate with this module using only the serial communication interface.

5.2 The Microcontroller Architecture

The User IO Unit was developed using M68HC11 microcontroller from Motorola Inc. This microcontroller which is commonly known as HC11 contains many built in peripherals such as SCI, SPI, RAM, EEPROM, ADC, GPIO and Timing modules. It is typically driven by an 8 MHz crystal oscillator, which provides a main clock speed of 2 MHz (the main clock rate divided by four). For LCD display and keypad handling purposes, this speed is sufficient. Shown in Figure 7 is the circuit diagram for the User IO Unit. Basically, the circuit is a minimum system that uses its GPIO pins to drive the necessary signals to the LCD module and decode the keypad.

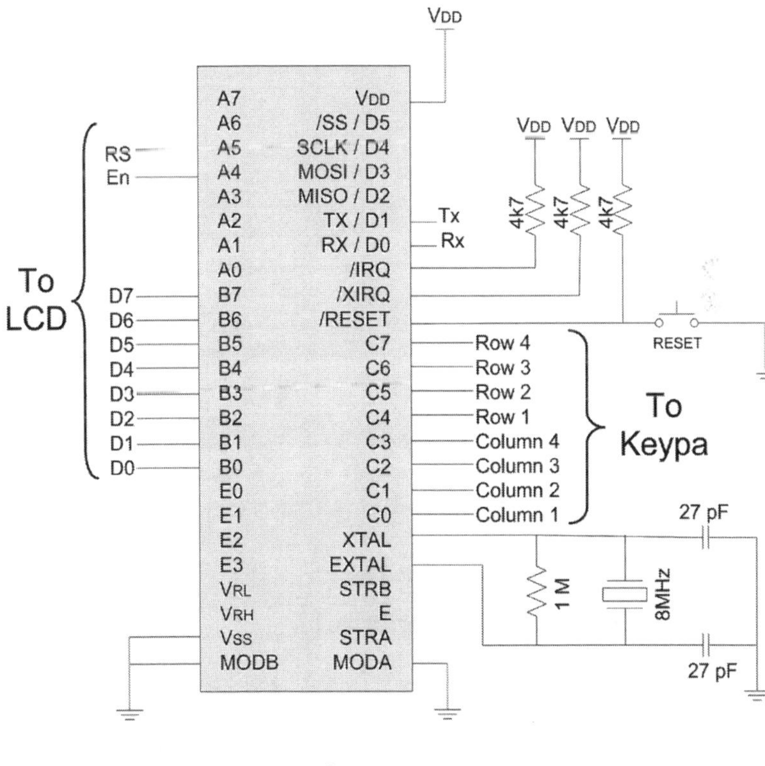

Figure 7. User IO Unit circuit diagram.

5.3. Liquid Crystal Display (LCD)

The Liquid Crystal Display is a standard 2 lines, 16 characters alphanumeric LCD module. The alphanumeric display is actually driven by a LCD driver built-in in the module. The LCD driver contains a table of standard alphanumeric characters stored in its volatile memory. Commands and data are sent to this driver by driving the necessary pins to certain patterns. Figure 8 shows the connections of the LCD module to the embedded system.

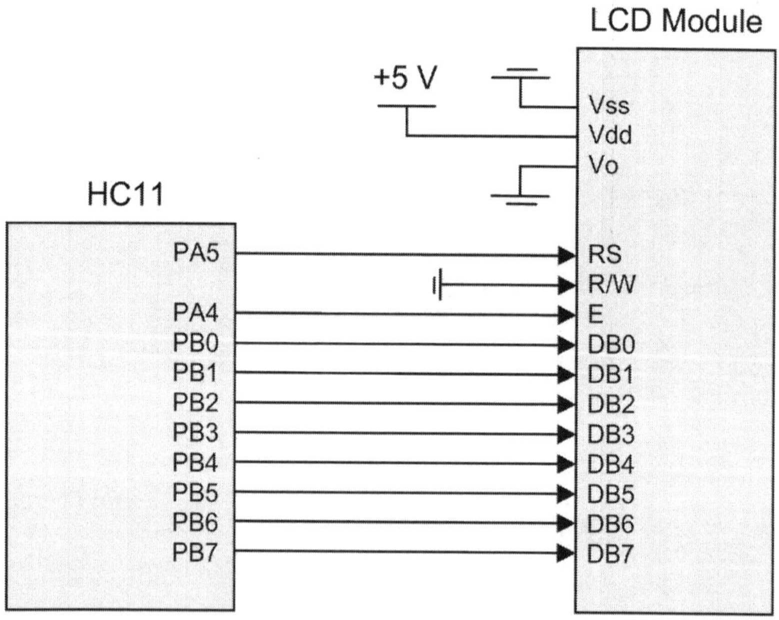

Figure 8. Interfacing LCD module with HC11.

In order to display data to the LCD module, standard protocol and commands set were used. These commands are usually standard among most LCD modules. Basically, the microcontroller needs to handle the control signals according to certain timing requirements and outputs the command or data to the data bus accordingly.

5.4. Keypad

The main usage of the keypad is as the input interface for the user when the PC GUI unit is not being used. Settings such as solid density and status of PC GUI unit (online or offline) are inputs that can be entered through this interface. The keypad used in this research is a standard 4×4 matrix keypad. This interface acts as an input to user to change settings if required. The microcontroller is responsible to decode the key pressed by the user and send it to the Main Controller Unit.

6. FIRMWARE DEVELOPMENT

6.1. System Flowchart

The overall operation performed by the DSP is described in Figure 9. Basically, the system was initialized at program start to set the configurations of ADC, SPI and SCI such as ADC channel sampling sequence, SCI and SPI bit length, parity and baud rate. After being initialized, data acquisition began by sending a pulse to Parallel Load pin on parallel to serial converters. This action latched the binary outputs from the comparator into the converter. Then, acquisition from rectilinear projection sensors was carried out by reading data from SPI. The use of SPI caused the binary values to be packed in six words, so they were unpacked into variables after the acquisition.

As for the orthogonal projections, the Select pins on the analogue multiplexers were set from 0 to 3 to multiplex 64 channels into the ADC. The Select pin setting and the respective channels being multiplexed are shown in Table 2 below.

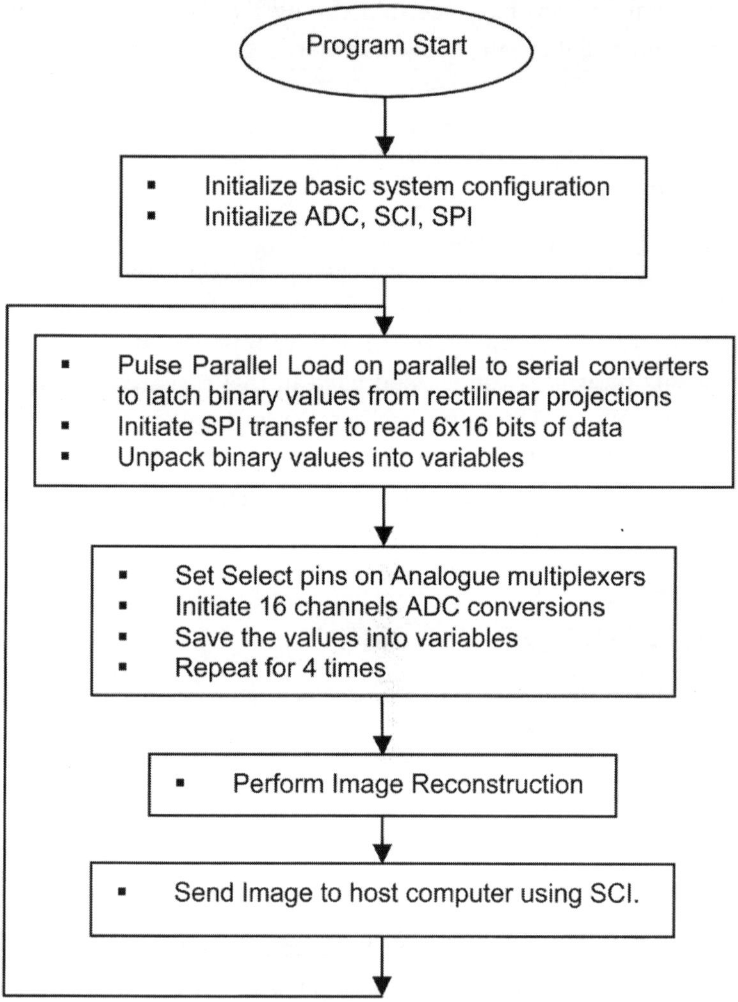

Figure 9. Overall System Flowchart.

Two image reconstruction algorithms were implemented in this study. The first algorithm was the commonly known Linear Back Projection algorithm while the second algorithm was the Hybrid algorithm [5]. The image reconstructed was stored in internal Random Access Memory (RAM) of the DSP. Finally, the DSP sent the image to the host computer upon request via SCI. At the host computer, a Graphical User Interface (GUI) was developed using Ms. Visual C++ to facilitate this feature.

Table 2. Analogue Multiplexers configuration.

Select pins		ADC channel							
		A0	*A1*	*A2*	*A3*	*A4*	*A5*	*A6*	*A7*
MSB	*LSB*	*B0*	*B1*	*B2*	*B3*	*B4*	*B5*	*B6*	*B7*
		Orthogonal Sensor Array index							
0	0	0	1	2	3	4	5	6	7
0	1	8	9	10	11	12	13	14	15
1	0	16	17	18	19	20	21	22	23
1	1	24	25	26	27	28	29	30	31

6.2. Hybrid Image Reconstruction Algorithm

In order to visualize the flow concentration using tomography technique, image reconstruction algorithm was applied to the measurement data. The Hybrid reconstruction algorithm which originated from the Linear Back Projection algorithm was implemented with some improvements. Basically, the back projection algorithm can be illustrated as in Figure 10.

Figure 10 (a) shows the actual object and light projection while Figure 10 (b) shows the image reconstructed using Linear Back Projection algorithm. Considering only a single object, this algorithm worked fine (only single pixel detected, shown by dark green) with some smearing, shown by light green. The values shown are the summation of the sensor voltage where sensor voltage is assumed to be 5V if blocked by object. Therefore, the detected pixel shows value of 10V. However, when there is more than one object, the Linear Back Projection algorithm results in ambiguous image as in Figure 11. Notice that there are four pixels 'detected' while there are only two actual objects. This is actually a result of summation of the smearing effects.

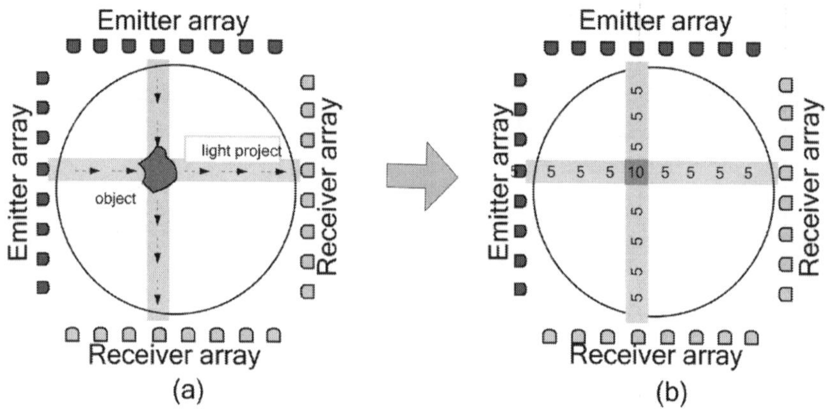

Figure 10. Back projection technique.

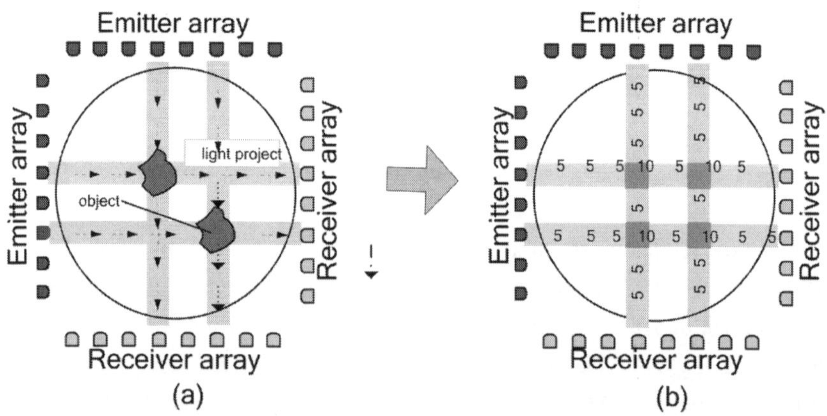

Figure 11. Ambiguous image resulting from back projection.

Based on work by Ibrahim *et al.* [4] ambiguous image can be eliminated by using at least four projections. He added two rectilinear projections to eliminate ambiguous image and named the image reconstruction algorithm as Hybrid Reconstruction algorithm. Pang [7] further improved this algorithm by doubling the resolution obtained from the previous research which used 8×8 sensor pairs in the orthogonal layer and 11×11 sensor pairs in the rectilinear layer [4] to 16×16 sensor pairs in the orthogonal layer and 23×23 sensor pairs in the rectilinear layer.

Basically, the additional number of sensors used in rectilinear layer was to project the masking layer's beam of light at the centre of each pixel, whereby object's presence at any pixel could be detected with only four sensor readings (two from orthogonal and two from rectilinear). In this research, the same hybrid image reconstruction algorithm was improved by further dividing each pixel into eight small triangles. This is shown in Figure 12.

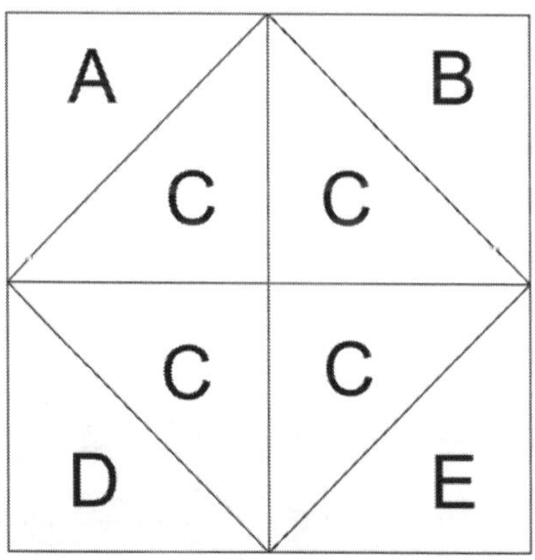

Figure 12. Division of a pixel into eight triangles.

Although the pixel was divided into 8 small triangles, four triangles in the centre were grouped as one since there wasn't any way to determine the smaller triangles' concentration independently based on the sensor readings. Thus, there were 5 parts, namely A, B, C, D and E. Object that is present at any of these 5 parts could be determined by back projecting the respective sensor readings. By dividing the pixel into 5 smaller parts, each of the small parts can be set as valid or invalid during concentration calculation. This differs from previous work by Pang where the whole pixel was set to a concentration value (using orthogonal layer values) upon confirming that an object exists at that pixel (using only 2 rectilinear layer values).

Using this approach, detection of arc edges of flow object can be presented more accurately. As a fair comparison, let's consider a pixel which is located at the centre of the flow region. The total area in this pixel is 28.3 mm^2 for the 85mm diameter pipe being used in this research. Therefore, each triangle contributes one-eight (⅛) of the total pixel, which is 3.5375 mm^2. Using this new Hybrid Reconstruction algorithm, there could be twelve possible cases to represent each pixel instead of only one case. These are shown in Figure 13, where triangle parts are shaded to represent blocked location.

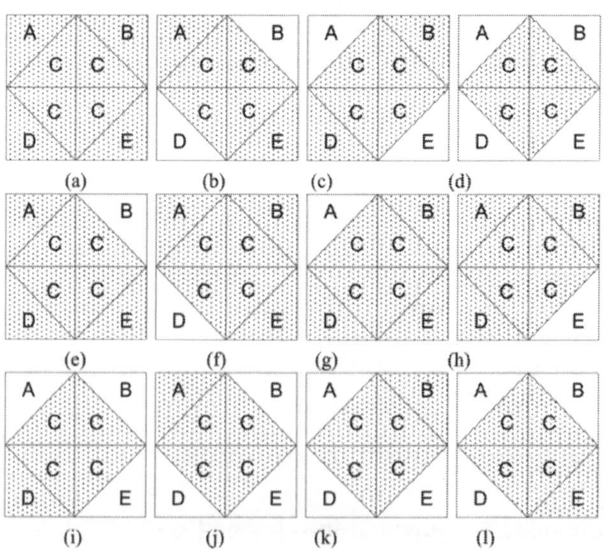

Figure 13. Different pixel reconstruction cases.

From these cases, the percentage of error can be calculated by comparing each case to case (a). This comparison is shown in Table 4.

Table 4. Comparison of pixel reconstruction accuracy.

Case	Assumption of Object	Area (Hybrid)	Area (new Hybrid)	% Error
a	A cube	28.3000	28.3000	0
b	A diagonally thin object	28.3000	21.2250	25
c	A diagonally thin object	28.3000	21.2250	25
d	A cylinder with small radius	28.3000	14.1500	50
e	An object's arc, top right	28.3000	24.7625	12.5
f	An object's arc, bottom left	28.3000	24.7625	12.5
g	An object's arc, top left	28.3000	24.7625	12.5
h	An object's arc, bottom right	28.3000	24.7625	12.5
i	A small particle with edge	28.3000	17.6875	37.5
h	A small particle with edge	28.3000	17.6875	37.5
k	A small particle with edge	28.3000	17.6875	37.5
l	A small particle with edge	28.3000	17.6875	37.5

Each triangle can be determined as valid or not by considering the following table. OFFSET_P and OFFSET_Q are constants for offset adjustments while iP and iQ are used for array indexing. Equation 1 and 2 shows the relationship between the array index with x-axis and y-axis.

where,

- y = y-axis location
- OFFSET_P = 11
- x = x-axis location
- OFFSET_Q = -4

VALID_RECT is a constant (0×80) being used to represent the binary bit value of TRUE from the rectilinear sensors acquisition. The value is set to 0 if the bit is FALSE. The following Table 5 shows the comparisons made in the software to determine the valid triangles using the New Hybrid Image Reconstruction algorithm. The valid triangle is marked with a tick sign when the comparison equations on the left are fulfilled.

Table 5. Comparison equations to determine valid triangle.

Comparison Equation	Valid Triangle				
	A	B	C	D	E
sensQ[iQ] = VALID_RECT sensP[iP+1] = VALID_RECT	✓				
sensQ[iQ] = VALID_RECT sensP[iP] = VALID_RECT			✓		
sensQ[iQ] = VALID_RECT sensP[iP-1] = VALID_RECT					✓
sensP[iP] = VALID_RECT sensQ[iQ+1] = VALID_RECT		✓			
sensP[iP] = VALID_RECT sensQ[iQ-1] = VALID_RECT				✓	

By using iP and iQ, the number of additions and subtractions carried out per pixel can be reduced compared to calculating it every time the comparison equations are evaluated. Besides that, the comparisons were carried out by evaluating comparison equations that involved iQ first. This prevents the Main Controller Unit from calculating iQ at each comparison. The same method was used for iP. Although it seemed that the reduction was very little, accumulation of the reductions for image reconstruction of more than a hundred sets of data made the reductions significant.

In previous work by Pang [7], all the cases shown in Figure 13 previously were considered as one case which was case (a). This was because the information from the masking layer was used only to eliminate the ambiguous image and did not contribute to the pixel value (the pixel value was calculated only from the orthogonal layer sensors). By using this New Hybrid Reconstruction algorithm, the information from the masking layer was fully utilized to determine the total value of a pixel. However, using this new algorithm required at least two bytes; one byte to keep the pixel value and the second byte to carry information on which area was valid (using bit or flag representation). Although this new algorithm brings more accuracy, it affects the overall system performance in a few areas. Firstly, the total data and therefore memory required to store the reconstructed images will be larger. Secondly, the data transfer to the host computer for reconstructed image display will be slower as more data needed to be sent. Thirdly, the amount of cross correlation operation required at later stage to calculate flow velocity will increase and this decreased the system performance.

Therefore, an alternative method was essentially required to represent the equivalent information and yet would not burden the system too much. Finally, this technique was implemented by trading off the actual graphical representation to an equivalent pixel value. This can be understood by realizing that no matter how the pixel concentration was represented by 5 areas; ultimately it would be converted to a single pixel value for mass flow rate measurement. The only difference was that the triangles were displayed as a full pixel with equivalent concentration value of total valid triangles instead of visual representation of actual detected independent triangles. In other words, it was a trade off of the valid triangle location for higher performance but still maintain the more accurate equivalent value of the concentration measurement.

In order to do this, all the triangles were summed up and averaged to represent the whole pixel. This way, the small triangles were taken into account to represent the whole pixel's value. Mathematically, each pixel's value was calculated according to the following equation.

$$V_{Pixel} = \frac{\sum Valid\ Triangles}{8} \times Pixel\ concentration \tag{4.4}$$

The image reconstruction algorithm used in this research is shown graphically in the flow chart diagram of Figure 14.

6.3. DSP Optimization Techniques

Programming a DSP is not as straight forward as programming a high level programming language. Although the DSP could be programmed using a C compiler, the code implementation must be well considered in order to optimize for speed. Using a personal computer nowadays with Giga Hertz operating frequency almost have no noticeable delay but for a DSP running at only 150 MHz, some delay here and there could cause very noticeable system slow down.

Among the techniques used to speed up the processing are:

1 Use internal memory to perform calculations such as Fast Fourier Transform to reduce wait states required to access external memory

2 Reduce the number of parameters to be passed through frequently called functions. This will reduce the time required to pass the extra parameters to the stack.

3 Use fixed point calculation instead of floating ('float' type). Real values requires conversion algorithm which takes longer time than the integer values.

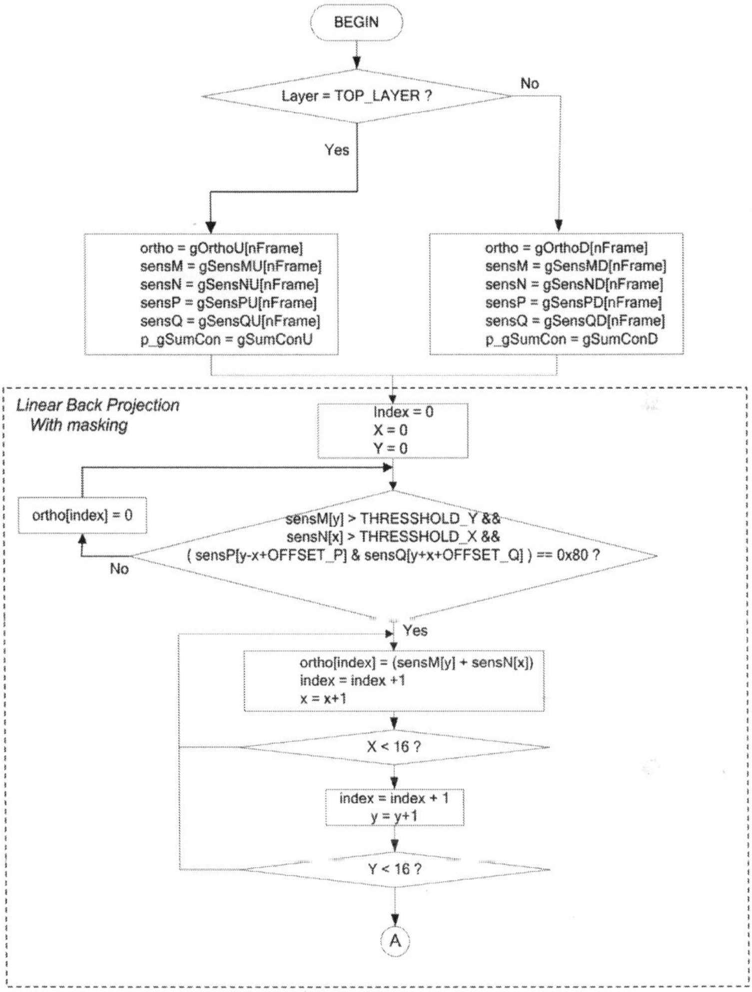
BEGIN
Layer = TOP_LAYER ?
No
Yes
ortho = gOrthoU[nFrame]
sensM = gSensMU[nFrame]
sensN = gSensNU[nFrame]
sensP = gSensPU[nFrame]
sensQ = gSensQU[nFrame]
p_gSumCon = gSumConU
ortho = gOrthoD[nFrame]
sensM = gSensMD[nFrame]
sensN = gSensND[nFrame]
sensP = gSensPD[nFrame]
sensQ = gSensQD[nFrame]
p_gSumCon = gSumConD
Linear Back Projection
With masking
Index = 0
X = 0
Y = 0
ortho[index] = 0
sensM[y] > THRESSHOLD_Y &&
sensN[x] > THRESSHOLD_X &&
(sensP[y-x+OFFSET_P] & sensQ[y+x+OFFSET_Q]) == 0x80 ?
No
Yes
ortho[index] = (sensM[y] + sensN[x])
index = index +1
x = x+1
X < 16 ?
index = index + 1
y = y+1
Y < 16 ?
A

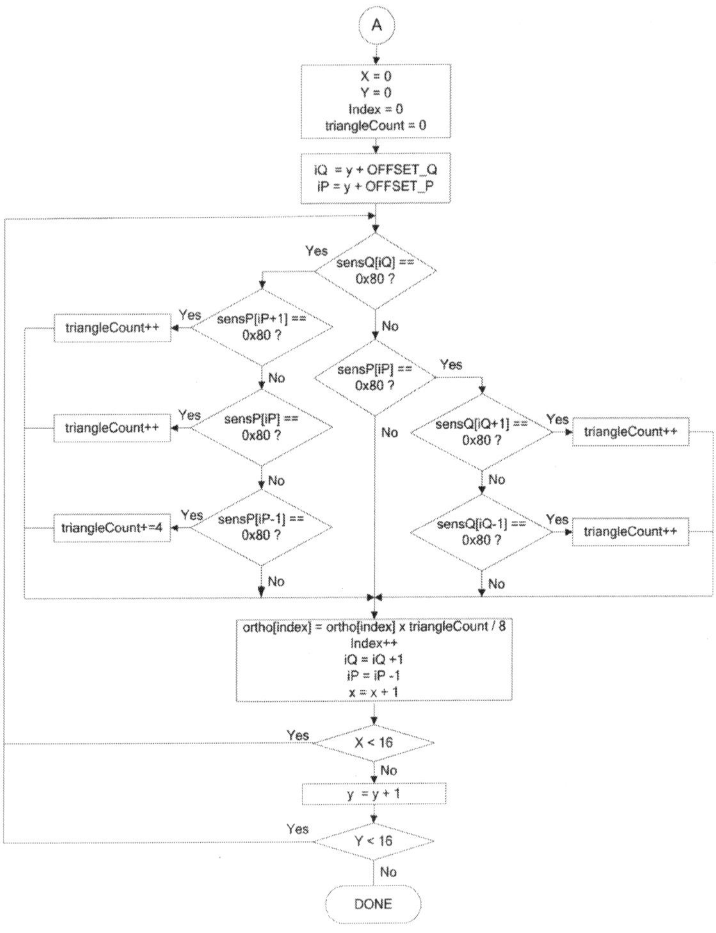

Figure 14. New Hybrid Reconstruction Algorithm.

7. RESULTS AND DISCUSSIONS

The results acquired were evaluated in two categories. The first category was the execution speed while the second category was the quality of the reconstructed cross sectional image.

7.1. The Execution Speed

The execution speed can be measured easily by toggling a General Purpose Input Output (GPIO) pin and measuring the time delay using an oscilloscope. The timer function available in the DSP could was also used to measure the execution time. From experiments, the measured time using timer DSP is the same as using oscilloscope. A GPIO pin was set to HIGH at the beginning of data acquisition and set to LOW to represent the beginning of image reconstruction. Therefore, the overall time per frame was measured from a HIGH level to the next HIGH level. The time required per frame for one layer and both layers are shown in Table 3 below.

Table 3. Program Execution Time.

Algorithm	One Layer	Both Layers
Linear Back Projection	233 us	313 us
Hybrid	305 us	466 us

The time required for data acquisition and processing is comparable to the time required in previous work by Pang [8] which needed 1.58 ms per frame. This improvement is mainly due to the use of high speed ADC core in the DSP, digital sampling of rectilinear projection sensors and high speed execution using DSP.

7.2. The Reconstructed Image

Two small sticks (6 mm diameter), medium sized pipe (2.7cm diameter) and a large sized pipe (4.24 cm diameter) were used as samples to be imaged. The results are shown in Figures 15, 16 and 17. All figures to show three things:

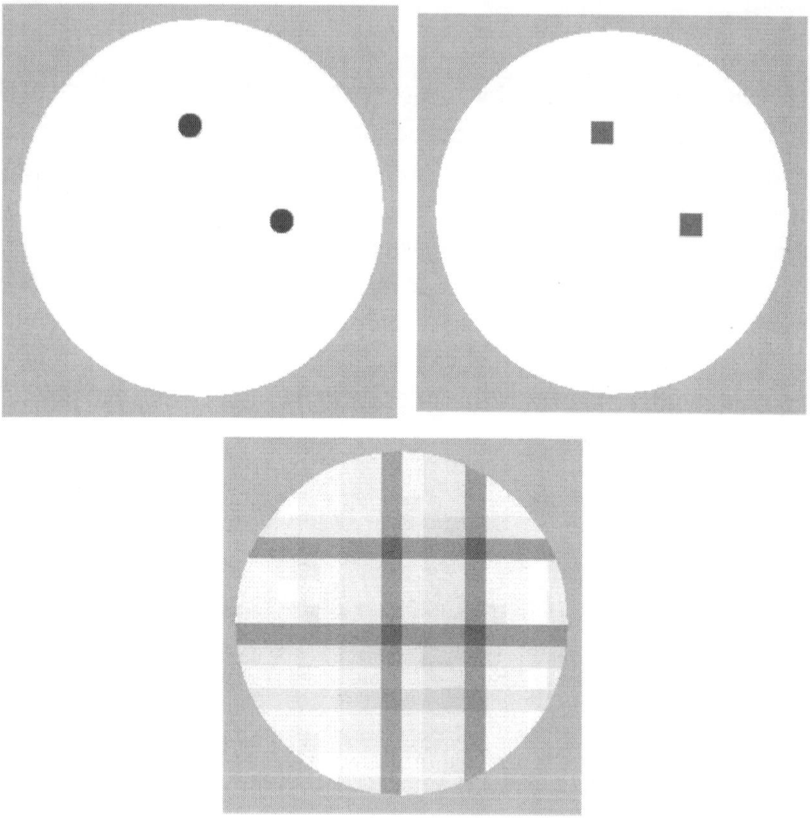

Figure 15. Figure 15a. Two small 6 mm diameter sticks.**Figure 15b.** The hybrid algorithm reconstructs an image showing two square blocks.**Figure 15c.** The LBP algorithm shows smearing and aliasing/ambiguity of the image.

a. The object being imaged

b. The hybrid LBP image

c. The simple LBP image

Results for three tests are presented. Figure 15a shows two small 6 mm diameter sticks. The hybrid algorithm reconstructs an image showing two square blocks (Figure 15b), but the LBP algorithm shows smearing and aliasing/ambiguity of the image (Figure 15c).

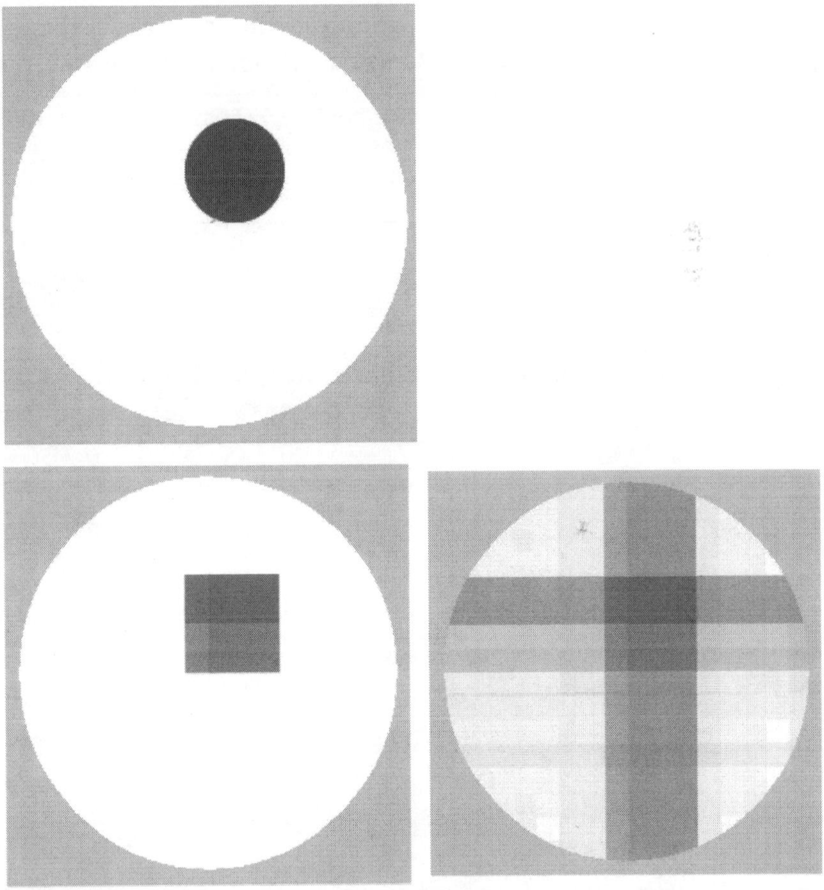

Figure 16. Figure 16a. The 2.7 cm diameter pipe.**Figure 16b.** The hybrid algorithm reconstructs an image showing a square.**Figure 16c.** The LBP algorithm shows smearing of the image.

Figure 16a shows a 2.7 cm diameter pipe. The hybrid algorithm reconstructsan image showing a square (Figure 16b), but the LBP algorithm shows smearing of the image (Figure 16c).

Figure 17a shows a 4.24 cm diameter pipe. The hybrid algorithm reconstructs an image showing an approximation to a circle (Figure 17b), but, in comparison, the LBP algorithm shows a badly smeared image.

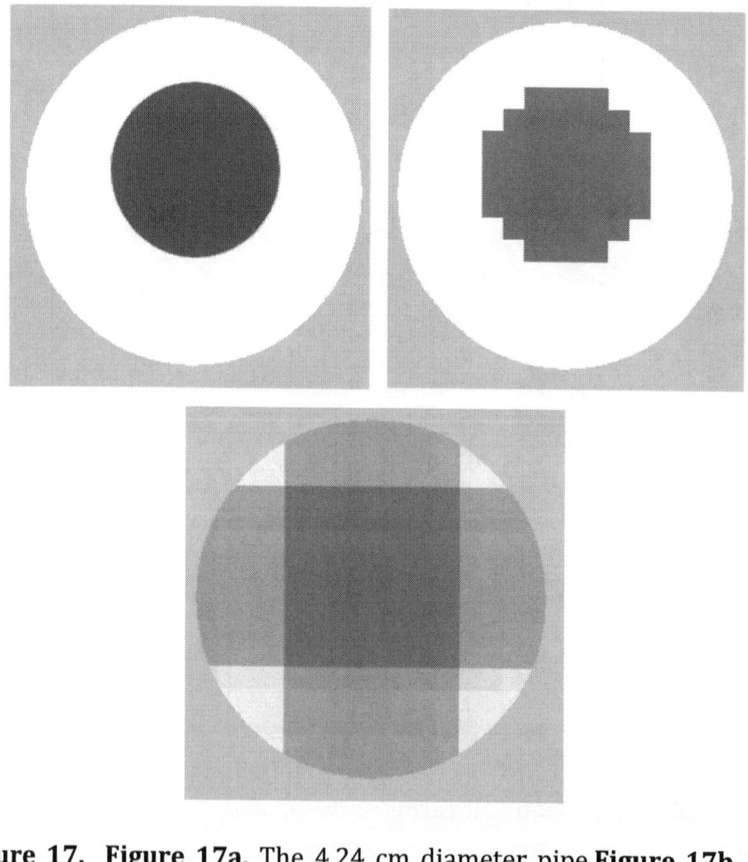

Figure 17. Figure 17a. The 4.24 cm diameter pipe.**Figure 17b.** The hybrid algorithm reconstructs an image showing an approximation to a circle.**Figure 17c.** The LBP algorithm shows a badly smeared image.

8. CONCLUSIONS

The data acquisition and processing of a parallel beam projection optical tomography system was presented and discussed in this paper. The speed at which the DSP perform data acquisition and image reconstruction is comparable to previous work utilizing DAQ and host computer. This proves that DSP is a promising solution at a lower price, smaller board size and yet competitive in performance.

REFERENCES

1. Beck, M.S.; Green, R.G.; Thorn, R. Non-intrusive measurement of solids mass flow in pneumatic conveying. *J. Phys. E: Sci. Instrum.* **1987**, *20*(7), 835–840.

2. Chan, K.S. Real time image reconstruction for fan beam optical tomography system. M.Sc. Thesis, Universiti Teknologi Malaysia, 2002.

3. Green, R.G.; Kwan, H.K.; John, R.; Beck, M.S. A low-cost solids flowmeter for industrial use. *J. Phys. E: Sci. Instrum.* **1978**, *11*(10), 1005–1010.

4. Ibrahim, S.; Green, R.G.; Evans, K.; Dutton, K.; Rahim, R.A. Optical tomography for process measurement and control. *UKACC International Conference* **2000**, 4–7.

5. Ibrahim, S.; Green, R.G. Velocity measurement using optical sensors. *Proceedings ICSE* **2002**, 126–129.

6. Ibrahim, S.; Green, R.G. Optical fibre sensors for imaging concentration profile. *Proceedings. ICSE* **2002**, 121–125.

7. Pang, J.F.; Rahim, R.A.; Chan, K.S. Infrared sensor configuration using four parallel beam projections. *Proceeding 3rdInternational Symposium on Process Tomography in Poland* **2004**, 48–51.

8. Pang, J.F.; Rahim, R.A.; Chan, K.S. Real Time Image Reconstruction System Using Two Data Processing Units in Optical Tomography. *Proceeding 3rd International Symposium on Process Tomography in Poland* **2004**, 52–55.

9. Rahim, R.A. A tomography imaging system for pneumatic conveyors using optical fibres. Ph. D. Thesis, Sheffield Hallam University, 1996.

10. Rahim, R.A.; Green, R.G.; Horbury, N.; Dickin, F.J.; Naylor, B.D.; Pridmore, T.P. Further development of a tomographic imaging system using optical fibres for pneumatic conveyors. *Meas. Sci. Technol.* **1996**, *7*(3), 419–422.

11. Texas Instruments. *DSP Selection Guide* **2002**.

12. Thiam, C.K.; Cong, T.Z.; Fea, P.J.; Rahim, R.A.; Loon, G.C. Digital sensors for masking purpose in parallel beam optical tomography system. *Proceeding 3rd International Symposium on Process Tomography in Poland* **2004**, 158–160.

13. Xie, C.G.; Stott, A.L.; Huang, S.M.; Plaskowski, A.; Beck, M. S. Mass-flow measurement of solids using electrodynamic and capacitance transducers. *J. Phys. E: Sci. Instrum.* **1989**, *22*(9), 712–719.

CHAPTER 8

A Real-Time Video Encoding Scheme Based on the Contourlet Transform

Stamos Katsigiannis[1], Georgios Papaioannou[2] and Dimitris Maroulis[1]

[1] Department of Informatics and Telecommunications, National and Kapodistrian University of Athens, Athens, Greece

[2] Department of Informatics, Athens University of Economics and Business, Athens, Greece

1. INTRODUCTION

Real-time video communication over the internet and other heterogeneous IP networks has become a significant part of modern communications, underlining the need for highly efficient video coding algorithms. The most desirable characteristic of such an algorithm would be the ability to maintain satisfactory visual quality while achieving good compression. Additional advantageous characteristics would be low computational complexity and real-time performance, allowing the algorithm to be used in a wide variety of less powerful computers. Transmission of video over the network would benefit by the ability to adapt to the network's end-to-end bandwidth and transmitter/receiver resources, as well as by resistance to packet losses that might occur. Additionally, scalability and resistance to noise would be highly advantageous characteristics for a modern video compression algorithm. Most state of the art video compression techniques like the H.264, DivX/Xvid, MPEG2 fail to achieve real time performance without the use of dedicated hardware due to their high computational complexity. Moreover, in order to achieve optimal compression and quality they depend on multipass statistical and structural analysis of the whole video content, which cannot happen in cases of live video stream generation as in the case of video-conferencing.

In this chapter, a more elaborate analysis of a novel algorithm for high-quality real-time video encoding, originally proposed in [1], is presented. The algorithm is designed for content obtained from low resolution sources like web cameras, surveillance cameras, etc. Critical to the efficiency of video encoding algorithm design is the selection of a suitable image representation method. Texture representation methods proposed in the literature that utilize the Fourier transform, the Discrete Cosine transform, the Wavelet transform as well as other frequency domain methods have been extensively used for image and video encoding. Nevertheless, these methods have some limitations that have been partially addressed by the Contourlet Transform (CT) [2], which our video encoding algorithm is based on. The Contourlet Transform offers multiscale and directional decomposition, providing anisotropy and directionality, features missing from traditional transforms like the Discrete Wavelet Transform [2]. In recent years, the Contourlet Transform has been successfully utilised in a variety of texture analysis applications, including synthetic aperture radar (SAR) [3], medical and natural image classification [4], image denoising [5], despeckling of images, image compression, etc. By harnessing the computational power offered by modern graphics processing units

(GPUs), a gpu-based contourlet transform is able to provide an image representation method with advantageous characteristics, while maintaining a fast performance.

The rest of this chapter is organised in four sections. First, some background knowledge and information needed for better understanding the algorithm is presented in section 2. Then, the aforementioned video encoding algorithm is presented in section 3, whereas an experimental study for the evaluation of the algorithm is provided in section 4. Conclusions and future perspectives of this work are presented in section 5.

2. BACKGROUND

2.1. The Contourlet Transform

The Contourlet Transform (CT) is a directional multiscale image representation scheme proposed by Do and Vetterli, which is effective in representing smooth contours in different directions of an image, thus providing directionality and anisotropy [2]. The method utilizes a double filter bank in which, first the Laplacian Pyramid (LP) [6] detects the point discontinuities of the image and then the Directional Filter Bank (DFB) [7] links those point discontinuities into linear structures. The LP provides a way to obtain multiscale decomposition. In each LP level, a downsampled lowpass version of the original image and a more detailed image with the supplementary high frequencies containing the point discontinuities are obtained. This scheme can be iterated continuously in the lowpass image and is restricted only by the size of the original image due to the downsampling. The DFB is a 2D directional filter bank that can achieve perfect reconstruction, which is an important characteristic for image and video encoding applications. The simplified DFB used for the contourlet transform consists of two stages and leads to 2^l subbands with wedge-shaped frequency partitioning [8], with l being the level of decomposition. The first stage of the DFB is a two-channel quincunx filter bank [9] with fan filters that divides the 2D spectrum into vertical and horizontal directions, while the second stage is a shearing operator that just reorders the samples. By adding a 45 degrees shearing operator and its inverse before and after a two-channel filter bank, a different directional frequency partition is obtained (diagonal directions), while maintaining the ability to perfectly reconstruct the original image, since the sampling locations coincide with the (integer) pixel grid.

The combination of the LP and the DFB is a double filter bank named Pyramidal Directional Filter Bank (PDFB). In order to capture the directional information, bandpass images from the LP decomposition are fed into a DFB. This scheme can be repeated on the coarser image levels. The combined result is the contourlet filter bank, which is a double iterated filter bank that decomposes images into directional subbands at multiple scales. The contourlet coefficients have a similarity with wavelet coefficients since most of them are almost zero and only few of them, located near the edge of the objects, have large magnitudes [10]. In the presented algorithm, the Cohen and Daubechies 9-7 filters [11] have been utilized for the Laplacian Pyramid. For the Directional Filter Bank, these filters were mapped into their corresponding 2D filters using the McClellan transform as proposed by Do and Vetterli in [2]. It must be noted that these filters are not considered as optimal. The creation of optimal filters for the contourlet filter bank remains an open research topic. An outline of the Contourlet Transform is presented on Figure 1, while an example of decomposition is shown on Figure 2.

2.2. GPU-based contourlet transform

By analysing the structure of the contourlet transform, it is evident that its most computationally intensive part is the calculation of all the 2D convolutions needed for complete decomposition or reconstruction. Calculating the convolutions on the CPU using the 2D convolution definition is not feasible for real-time applications since performance suffers significantly due to the computational complexity. Utilizing the DFT or the FFT in order to achieve better performance provides significantly faster implementations but still fails to achieve satisfactory real-time performance, especially in mobile platforms such as laptops and tablet PCs. The benefits of the FFT for the calculation of 2D convolution can only be fully exploited by an architecture supporting parallel computations. Modern personal computers are commonly equipped with powerful graphics processors (GPUs), which in the case of live video capture from web or surveillance cameras are underutilized. Intensive, repetitive computations that can be computed in parallel can be accelerated by harnessing this "dormant" computational power. General purpose computing on graphics processing units (GPGPU) is the set of techniques that use a GPU, which is otherwise specialized in handling computations for the display of computer graphics, in order to perform computations traditionally handled by a CPU. The highly parallel structure of GPUs makes them more effective than general-purpose CPUs for algorithms where processing of large blocks of data can be done in parallel.

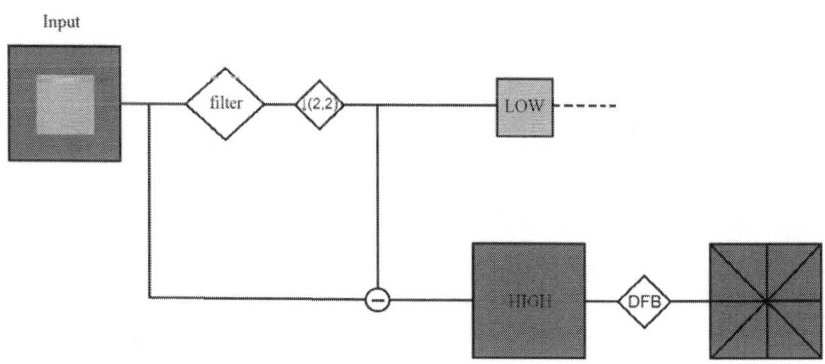

Figure 1. The Contourlet Filter Bank.

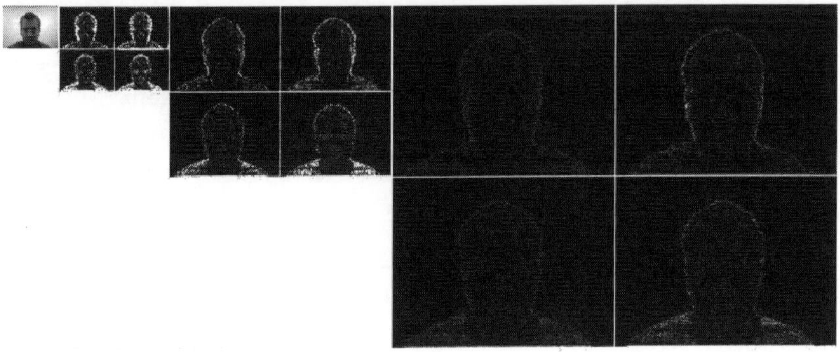

Figure 2. Example of contourlet transform decomposition of a greyscale image. Three levels of decomposition with the Laplacian Pyramid were applied, each then decomposed into four directional subbands using the Directional Filter Bank.

For the GPU implementation of the contourlet transform, the NVIDIA Compute Unified Device Architecture (CUDA) has been selected due to the extensive capabilities and specialized API it offers. CUDA is a general purpose parallel computing architecture that allows the parallel compute engine in NVIDIA GPUs to be used in order to solve complex computational problems that are outside the scope of graphics algorithms. In order to compute the contourlet transform, first the image and the filters are transferred from the main memory to the GPU dedicated memory. Then, the contourlet transform of the image is calculated by migrating all the calculations on the GPU in order to reduce the unnecessary transfers to and from the main memory that introduce delay to the computations. The 2D convolutions required are calculated by means of the FFT. After calculating the contourlet transform of the image, the output is transferred back to the main memory and the GPU memory is freed. Considering that this implementation will be used for video encoding, the filters are loaded once at the GPU memory since they will not change from frame to frame. In order to evaluate the performance of this approach, various implementations of the contourlet transform were developed, both for the CPU and the GPU. These implementations were based on the FFT (frequency domain) and the 2D convolution definition (spatial domain). Except for the basic GPU implementation using spatial domain convolution, other out-of-core implementations were developed, based on the 2D convolution definition and utilizing memory management schemes in order to support larger frames when the GPU memory is not sufficient [12]. The GPU implementation based on the FFT outperformed all the aforementioned implementations in our tests and was therefore the method of choice for our video encoder.

2.3. The YCoCg colour space

Inspired by recent work on real-time RGB frame buffer compression using chrominance subsampling based on the YCoCg colour transform [13], we investigated the use of these techniques in conjunction with the contourlet transform to efficiently encode colour video frames.

The human visual system is significantly more sensitive to variations of luminance compared to variations of chrominance. Encoding the luminance channel of an image with higher accuracy than the chrominance channels provides a simple low complexity compression scheme, while maintaining satisfactory visual quality. Various image and video compression algorithms take advantage of this fact in order to achieve increased efficiency. First introduced in H.264 compression, the RGB to YCoCg transform decomposes a colour image into luminance (Y), orange chrominance (Co) and green chrominance (Cg) components and has been shown to exhibit better decorrelation properties than YCbCr and similar transforms [14]. It was developed primarily to address some limitations of the different YCbCr colour spaces [15]. The transform and its reverse are calculated by the following equations:

$$Y = R/4 + G/2 + B/4 \tag{1}$$

$$Co = R/2 - B/2 \tag{2}$$

$$Cg = -R/4 + G/2 - B/4 \tag{3}$$

$$R = Y + Co - Cg \tag{4}$$

$$G = Y + Cg \tag{5}$$

$$B = Y - Co - Cg \tag{6}$$

Table 1. Average PSNR obtained for each image set after transforming from RGB to YCoCg and back using the same precision for the RGB and YCoCg components.

Image set	Number of images	Average PSNR (dB)
Kodak	23	59.27
Canon	18	59.05
Outdoor scene images	963	58.87

In order for the reverse transform to be perfect and to avoid rounding errors, the Co and Cg components should be stored with higher precision than the RGB components. Experiments using 23 images from the Kodak image set and 18 images from the Canon image set, all obtained from [16], as well as 963 outdoor scene images obtained from [17], showed that using the same precision for the YCoCg and RGB data when transforming from RGB to YCoCg and back results in an average PSNR of more than 58.87 dB for all the image sets, as shown in Table 1. This loss of quality cannot be perceived by the human visual system, resulting to no visible alteration of the image. Nevertheless, it indicates the highest quality possible when used for image compression.

3. THE PRESENTED ALGORITHM

Listings 1 and 2 depict the presented algorithm for encoding and decoding respectively. Input frames are considered to be in the RGB format. The first step of the algorithm is the conversion from RGB to YCoCg colour space for further manipulation of the luminance and chrominance channels. The luminance channel is decomposed using the contourlet transform, while chrominance channels are subsampled by a user-defined factor *N*. The levels and filters for contourlet transform decomposition are also defined by the user. From the contourlet coefficients obtained by decomposing the luminance channel, only a user-specified percentage of the most significant ones are retained. Then, the precision allocated for storing the contourlet coefficients is reduced. All computations up to this stage are performed on the GPU, avoiding unnecessary memory transfers from the main memory to the GPU memory and vice versa. After reducing the precision of the retained contourlet coefficients of the luminance channel, the directional subbands are encoded using a run length encoding scheme that encodes only zero valued elements. The large sequences of zero-valued contourlet coefficients that occur after the insignificant coefficient truncation make run length encoding ideal for their encoding.

The algorithm divides the video frames into two categories; keyframes and internal frames. Keyframes are frames that are encoded using the steps described in the previous paragraph and internal frames are the frames between two key frames. The interval between two keyframes is a user defined parameter. At the step before the run-length encoding, when a frame is identified as an internal frame all its components are calculated as the difference between the respective components of the frame and those of the previous key frame. This step is processed on the GPU while all the remaining steps of the algorithm are performed on the

CPU unless otherwise stated. Then, run length encoding is applied to the chromatic channels, the low frequency contourlet component of the luminance channel, as well as the directional subbands of the luminance channel. It must be noted that steps that are executed on the CPU are inherently serial and cannot be efficiently mapped to a GPU.

Listing 1. Steps of the encoding algorithm. Highlighted steps refer to calculations performed on the GPU, while the other steps refer to calculations performed on the CPU.

The encoding algorithm
Start
Input RGB frame
Convert to YCoCg
Downsample Co and Cg by *N*
Decompose Y with the Contourlet Transform
Keep the M% most significant CT coefficients
Round the CT coefficients to the *n*-th decimal
IF the frame is an internal frame
Calculate the frame as the difference between the frame and the previous keyframe
Run-length encoding of Co, Cg and the lowpass CT component of Y
END IF
Run-length encoding of the directional subbands of Y
Adjust precision of all components
IF frame is NOT the last frame
GOTO Start
END IF
Finish

The last stage of the algorithm consists of the selection of the optimal precision for each video component. The user can select between lossless or lossy change of precision, directly affecting the output's visual quality.

Listing 2. Steps of the decoding algorithm. Highlighted steps refer to calculations performed on the GPU, while the other steps refer to calculations performed on the CPU.

The decoding algorithm

Start
Input encoded frame
IF the frame is a keyframe
    Decode the run-length encoded directional subbands of Y
    Keep keyframe in memory and discard old keyframe
ELSE IF the frame is an internal frame
    Decode the run-length encoded Co, Cg, lowpass CT component of Y
    Decode the run-length encoded directional subbands of Y
    Calculate the frame as the sum of the frame and the previous keyframe
END IF
Upsample Co and Cg by N
Reconstruct Y
Convert to RGB
IF frame is NOT the last frame
    GOTO Start
END IF
Finish

3.1. Chrominance channel subsampling

Exploiting the fact that the human visual system is relatively insensitive to chrominance variations, in order to achieve compression, the chrominance channels Co and Cg are subsampled by a user-defined factor N that directly affects the output's visual quality and the compression achieved. The chrominance channels are stored in lower resolution, thus providing compression. For the reconstruction of the chrominance channels at the decoding stage, the missing chrominance values are replaced with the nearest available subsampled chrominance values. This approach is simple and naïve but has been selected due to the significantly smaller number of (costly) memory fetches and minimal computation cost, compared to other methods like bilinear interpolation. Utilizing the nearest neighbour reconstruction approach can introduce artifacts in the form of mosaic patterns in regions with strong chrominance transitions depending on the subsampling factor. In order to address this problem, given adequate computational resources, the receiver can choose to use the bilinear interpolation approach. Figure 3 shows an example of subsampling the Co and Cg chrominance channels by various factors, while using the nearest neighbour and the bilinear interpolation approach for reconstruction. Only a small, magnified part of the "baboon" image used is shown for clarity. As demonstrated in Figure 3, subsampling by a factor of 2 or 4 does not have a drastic effect on visual quality. Further subsampling leads to visible artifacts indicating the need for an optimal trade-off between quality and compression.

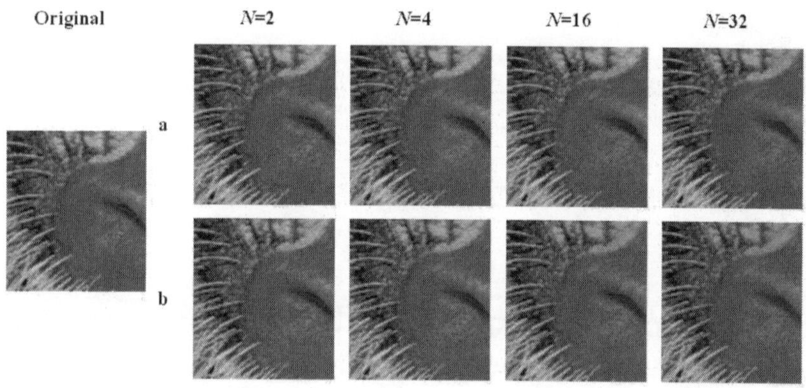

Figure 3. Example of chroma subsampling by factor *N* of the Co and Cg channels of the "baboon" image. Row (a) depicts images reconstructed using the nearest neighbour method, while (b) those reconstructed using bilinear interpolation.

3.2. Contourlet Transform decomposition of luminance channel and quality selection

The luminance channel of the frame is decomposed using the contourlet transform. Decomposition levels, as well as the filters used are user-defined and directly affect the quality of the output. Decomposition at multiple scales offers better compression while providing scalability, i.e. multiple resolutions inside the same video stream. This characteristic allows video coding algorithms to adapt to the network's end-to-end bandwidth and transmitter/receiver resources. The quality for each receiver can be adjusted without re-encoding the video frames at the source, by just dropping the encoded information referring to higher resolution than needed.

In order to achieve compression, after the decomposition of the luminance channel with the contourlet transform, a user-defined amount of the contourlet coefficients from the directional subbands are dropped by means of keeping only the most significant coefficients. The amount of coefficients dropped drastically affects the output's visual quality as well as the compression ratio. Contourlet coefficients with large magnitudes are considered more significant than coefficients with smaller magnitudes. Exploiting this fact, a common method for selecting the most significant contourlet coefficients is to keep the M most significant coefficients, or respective percentage, while dropping all the others [2] (coefficient truncation). This procedure leads to a large number of zero-valued sequences inside the elements of the directional subbands, a fact exploited by using run length encoding in order to achieve even higher compression. Considering the values and the distribution of contourlet coefficients at the directional subbands, only the zero-valued coefficients are run length encoded along the horizontal direction. Compression gained by run length encoding of all the different values is minimum and does not justify the increased computational cost. It is worth mentioning that dropping all the contourlet coefficients is similar to lowering the luminance channel's resolution while applying a lowpass filter and then, at the decoding stage, upscaling it without reincorporating the high frequency content.

Keeping only the most significant contourlet coefficients also provides a means to suppress the noise induced by low-quality sensors usually encountered in web-cameras. Random noise is largely unstructured and therefore not likely to generate significant contourlet coefficients [2]. As a result, keeping only the most significant contourlet coefficients provides enhanced visual quality, which is a highly desirable characteristic since no additional filtering of the video stream is required in order to reduce the noise level. On Figure 4, an example of smoothing due to the dropping of contourlet coefficients is shown. Mosaicing artifacts and noise introduced due to the low quality of the web camera's sensor are suppressed and replaced by a fuzzier texture, resulting in a smoother and more perceptually acceptable image.

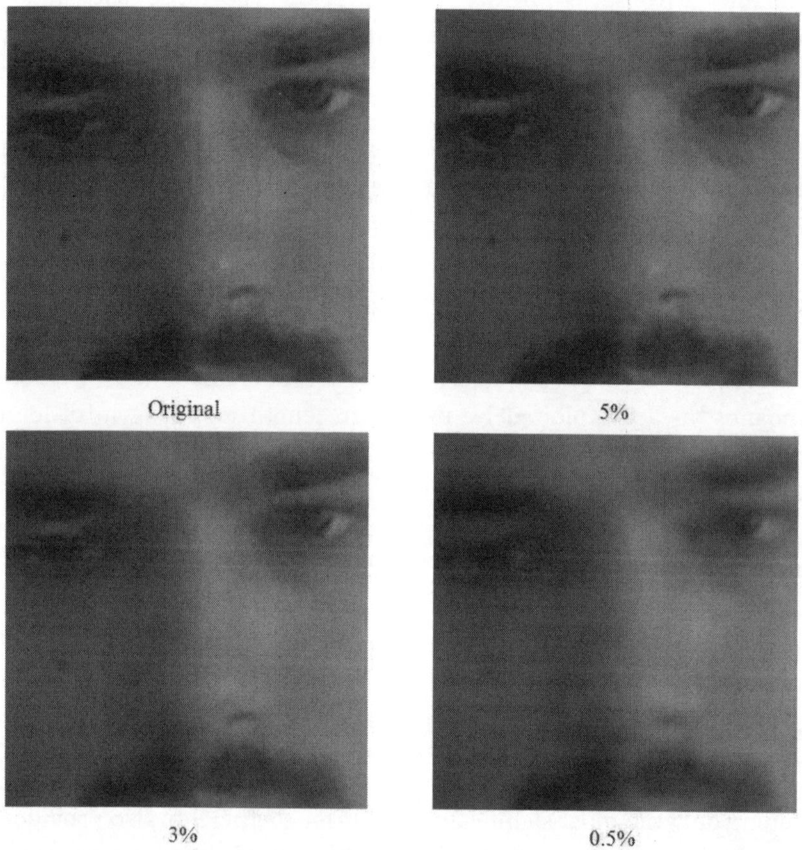

Figure 4. Example of smoothing due to the dropping of contourlet coefficients. The caption indicates the percentage of the contourlet coefficients retained. Images are cropped and scaled to 200% of their original size.

At the contourlet transform decomposition stage, 32 bit single precision floating point elements are used in order to avoid rounding errors and precision loss. Experiments with the precision allocated for the contourlet coefficients showed that the contourlet transform exhibits resistance to quality loss due to arithmetic precision reduction. This fact is exploited in order to achieve better compression by reducing the precision of the contourlet coefficients through rounding to a specific decimal point. Visual quality is not affected at all when only one decimal or more are kept. Rounding to the integer provides a PSNR of more than 60 dB when only the directional subbands' coefficients are rounded. Additionally, also rounding the low pass content provides a PSNR of more than 55 dB. In both cases, the loss of quality cannot be perceived by the human visual system and is considered as insignificant. For these experiments, the images were first transformed into the YCoCg colour space. Then the luminance channel was decomposed using the contourlet transform and the contourlet coefficients were rounded. No alteration was done to the chrominance channels. After the manipulation of the contourlet coefficients, the luminance channel was reconstructed and the image was transformed back into the RGB colour space.

3.3. Frame types

As mentioned before, frames are divided into keyframes and internal frames, with an internal frame being the difference between the current frame and the respective keyframe. Consecutive frames tend to have small variations, with many identical regions. This fact can be exploited by calculating the difference between a frame and the keyframe. This procedure provides components with large sequences of zero values leading to improved compression through the run length encoding stage. Especially in the case of video-conferencing or surveillance video, the background tends to be static, with slight or no variations at all. The occurrence of static background leads to many parts of the consecutive frames to be identical. As a result, calculating the difference of each frame from its respective keyframe provides large sequences of zero values leading to improved compression when run length encoding is applied. Run length encoding of the difference of contourlet-transformed images is even more efficient, since static noise is drastically suppressed by the coefficient truncation. Experiments showed that the optimal compression is achieved for a relatively small interval between keyframes, in the region of 5-7 internal frames, providing small groups of pictures (GOP) that depend to a keyframe. This characteristic makes the algorithm more resistant to packet loses when transmitting over a network. In the case of a scene change, consecutive frames drastically differ from each other and the compression achieved for the internal frames until the next keyframe is similar to that of a keyframe. If this scenario occurs, having small intervals between consecutive keyframes reduces the number of non optimally encoded frames. Nevertheless, in cases where the video is expected to be mostly static, like surveillance video for example, a larger interval between keyframes will provide considerably better compression.

3.4. Other supported colour spaces

Except for the YCoCg colour space, the algorithm supports the YCbCr and Greyscale colour spaces without the need to alter its core functionality. The process of encoding greyscale videos consists of handling the video as a colour video with only the luminance channel. All the steps of the algorithm are calculated except for those referring to the chromatic channels. On the other hand, due to the similarity of the YCbCr colour space with the YCoCg colour space the algorithm remains the same. The only difference is the RGB-to-YCbCr conversion at the encoder and the YCbCr-to-RGB conversion at the decoder. The luminance channel is identically handled as in the YCoCg-based algorithm, and the same holds for the replacement of CoCg channels with the CbCr ones. Nevertheless, the CbCr channels have a different range of values compared to CoCg channels. As a consequence, the optimal precision for the CbCr channels differs from that of the CoCg channels and has to be taken into consideration.

4. QUALITY AND PERFORMANCE ANALYSIS

For evaluating the presented algorithm, two videos were captured using a VGA web camera that supported a maximum resolution of 640x480 pixels. Low resolution web cameras are very common on everyday personal computer systems showcasing the need to design video encoding algorithms that take into consideration the problems arising due to low-quality sensors. The videos captured were a typical video-conference sequence with static background showing the upper part of the human body and containing some motion, and a surveillance video with almost no motion depicting the entrance of a building.

The captured videos were encoded using the YCoCg, YCbCr and Greyscale colour spaces. The chrominance channels of the colour videos were subsampled by a factor of 4 and the video stream contained two resolutions: the original VGA (640x480) as well as the lower QVGA (320x240). The method utilized for the reconstruction of the chrominance channels was the nearest neighbour method. The percentage of the most significant contourlet coefficients of the luminance channel retained was adjusted for each encoded video, providing results of various quality and compression levels. Furthermore, at each scale, the luminance channel's high frequency content was decomposed into four directional subbands. In order to test the algorithm using the YCbCr colour space, the RGB to YCbCr conversion formula for SDTV found in [18] was utilised. For the Greyscale colour space, the aforementioned videos were converted from RGB to greyscale using the standard NTSC conversion formula [18] that is used for calculating the effective luminance of a pixel:

$$Y(i, j)=0.2989 \cdot R(i, j) + 0.5870 \cdot G(i, j) + 0.1140 \cdot B(i, j) \tag{7}$$

The sample videos were encoded using a variety of parameters. The mean PSNR value for each video was calculated based on a set of different percentages of contourlet coefficients to be retained. The compression ratios achieved when using the scheme that incorporates both key frames and internal frames and when compressing all the frames as keyframes were also calculated. The interval between the key frames was set to five frames for the video-conference sample video and to twenty frames for the surveillance video. Detailed results are shown on Tables 2, 3 and 4 while sample frames of the encoded videos utilizing the YCoCg, YCbCr and Greyscale colour space for a set of settings are shown on Figures 5-10.

Examining the compression ratios achieved, it is shown that utilizing the keyframe and internal frame scheme outperforms the naive method of encoding all the frames the same way, as expected. However, the selection of an efficient entropy encoding algorithm that will further enhance the compression ability of our algorithm is still an open issue. Another interesting observation is that the contourlet transform exhibits substantial resistance to the loss of contourlet coefficients. Even when only 5% of its original coefficients are retained, the visual quality of the image is not seriously affected. This fact underlines the efficiency of the contourlet transform in approximating natural images using a small number of descriptors and justifies its utilization in this algorithm. The slightly lower PSNR achieved for the surveillance video sample can be explained due to the higher complexity of the scene compared to the video conference sample. More complex scenes contain higher frequency content, a portion of which is then discarded by dropping the contourlet coefficients.

Table 2. PSNRs achieved for the (a) video conference and (b) video surveillance samples, retaining various percentages of contourlet coefficients and utilizing the YCoCg. YCbCr and Greyscale colour spaces.

(a) Video conference sample			
	PSNR (dB)		
Contourlet coefficients retained (%)	YCoCg	YCbCr	Greyscale
10	45.11	44.77	52.04
5	44.53	44.29	49.71
3	43.88	43.70	47.72
1	42.30	42.23	44.28
0.5	41.62	41.56	43.10
0.2	41.30	41.25	42.60
0	39.15	39.13	39.82

(b) Video surveillance sample			
	PSNR (dB)		
Contourlet coefficients retained (%)	YCoCg	YCbCr	Greyscale
10	44.18	44.03	50.11
5	43.54	43.45	47.88
3	42.96	42.89	46.33
1	41.57	41.50	43.45
0.5	40.80	40.76	42.17
0.2	40.17	40.14	41.21
0	39.59	39.55	40.39

Table 3. Compression ratios achieved for the video conference sample, retaining various percentages of contourlet coefficients and utilizing the YCoCg. YCbCr and Greyscale colour spaces.

Video conference sample						
	Compression ratio					
Contourlet coefficients retained (%)	YCoCg		YCbCr		Greyscale	
	Only keyframes	Keyframes & Internal frames	Only keyframes	Keyframes & Internal frames	Only keyframes	Keyframes & Internal frames
10	4.96:1	11.06:1	4.93:1	12.05:1	2.09:1	3.49:1
5	6.44:1	14.39:1	6.44:1	16.31:1	2.94:1	4.44:1
3	7.36:1	16.39:1	7.37:1	18.98:1	3.56:1	5.02:1
1	8.71:1	19.46:1	8.70:1	23.15:1	4.57:1	5.85:1
0.5	9.07:1	20.24:1	9.06:1	24.33:1	4.87:1	6.07:1
0.2	9.22:1	20.62:1	9.21:1	24.81:1	5.00:1	6.17:1
0	11.71:1	25.84:1	11.71:1	32.89:1	7.65:1	7.53:1

Table 4. Compression ratios achieved for the video surveillance sample, retaining various percentages of contourlet coefficients and utilizing the YCoCg. YCbCr and Greyscale colour spaces.

Video surveillance sample						
	Compression ratio					
Contourlet coefficients retained (%)	YCoCg		YCbCr		Greyscale	
	Only keyframes	Keyframes & Internal frames	Only keyframes	Keyframes & Internal frames	Only keyframes	Keyframes & Internal frames
10	4.35:1	21.55:1	4.35:1	22.73:1	1.78:1	7.26:1
5	5.89:1	28.74:1	5.92:1	31.06:1	2.63:1	9.68:1
3	6.85:1	32.89:1	6.89:1	35.97:1	3.23:1	11.09:1
1	8.14:1	38.02:1	8.18:1	42.19:1	4.15:1	12.79:1
0.5	8.58:1	39.53:1	8.61:1	44.05:1	4.48:1	13.30:1
0.2	8.89:1	40.65:1	8.92:1	45.45:1	4.74:1	13.70:1
0	11.71:1	49.26:1	11.71:1	56.50:1	7.65:1	16.58:1

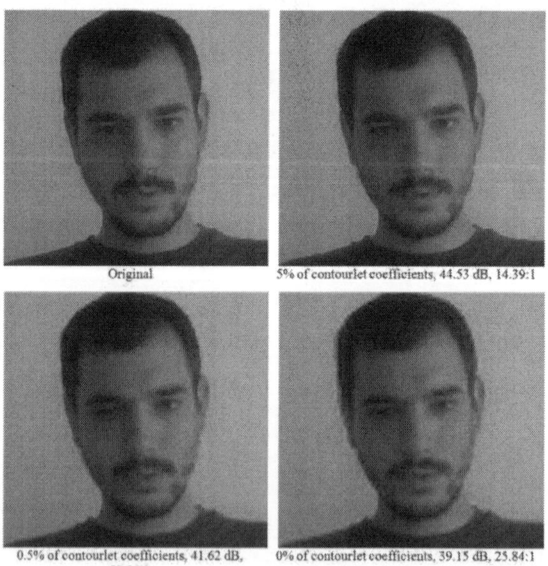

Figure 5. Sample frame of the encoded video-conference video for each setting using the *YCoCg* colour space. The frame has been resized and cropped to fit the figure.

Considering the YCoCg and the YCbCr colour spaces, for the two video samples tested, it is shown on Tables 2-4 and Figures 11 and 12 that the YCoCg colour space achieves slightly better visual quality (higher PSNR), while the YCbCr colour space provides better compression (higher compression ratio). The Greyscale examples cannot be directly compared to the colour samples since the calculated PSNR characterizes the original and encoded greyscale samples. Nevertheless, it is clear that in the case of Greyscale colour space, compression suffers greatly compared to the other colour spaces.

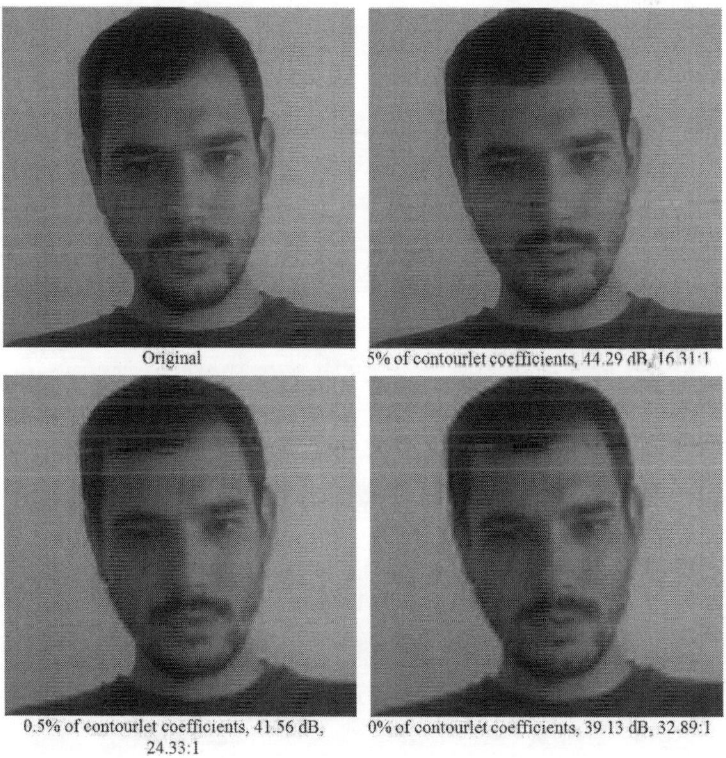

Figure 6. Sample frame of the encoded video-conference video for each setting using the *YCbCr* colour space. The frame has been resized and cropped to fit the figure.

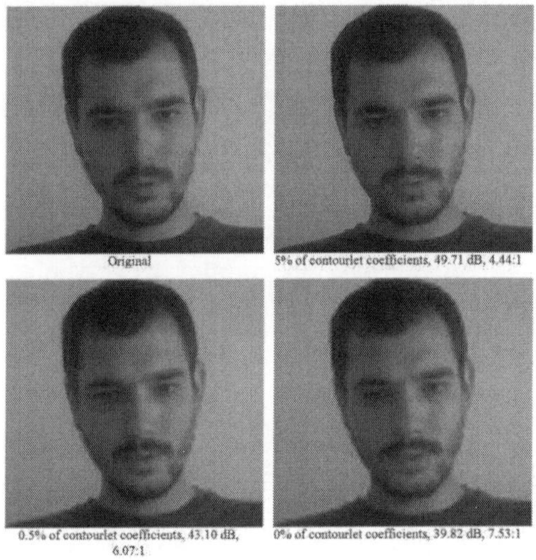

Figure 7. Sample frame of the encoded video-conference video for each setting using the *Greyscale* colour space. The frame has been resized and cropped to fit the figure.

Figure 8. Sample frame of the encoded video surveillance video for each setting using the *YCoCg* colour space. The frame has been resized and cropped to fit the figure.

Figure 9. Sample frame of the encoded video surveillance video for each setting using the *YCbCr* colour space. The frame has been resized and cropped to fit the figure.

Figure 10. Sample frame of the encoded video surveillance video for each setting using the *Greyscale* colour space. The frame has been resized and cropped to fit the figure.

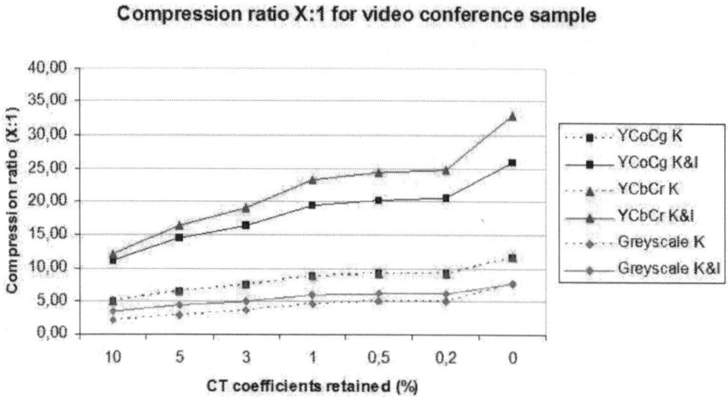

Figure 11. Compression ratio vs percentage of contourlet coefficients retained diagram, for the video conference sample, utilizing the YCoCg. YCbCr and Greyscale colour spaces. K refers to using only keyframes, while K&I refers to using both keyframes and internal frames.

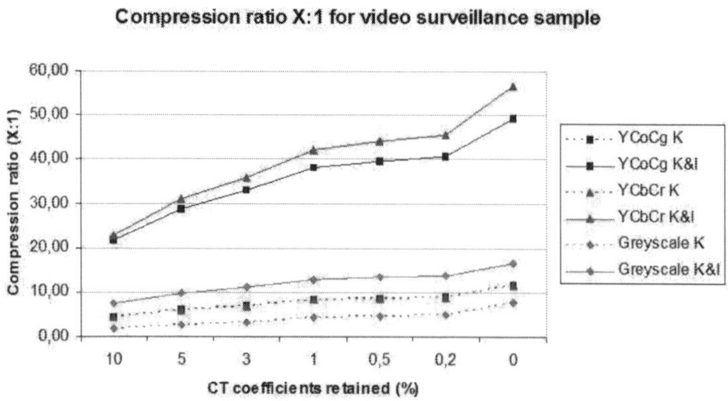

Figure 12. Compression ratio vs percentage of contourlet coefficients retained diagram, for the video surveillance sample, utilizing the YCoCg. YCbCr and Greyscale colour spaces. K refers to using only keyframes, while K&I refers to using both keyframes and internal frames.

Average execution times for the basic operations of the encoding and decoding algorithm for a frame of the video conference sample are presented on Table 5. Parameters were kept the same as in the previous examples and the computer utilised for the performance tests was equipped with an Intel Core i3 CPU, 4 GB of memory and a NVIDIA GeForce 430 graphics card with 1 GB of memory.

Table 5. Average execution times (in milliseconds) for the basic operations of the algorithm for a 640x480 video frame. The chrominance channels were subsampled by a factor of 4 and the video stream contained the original VGA (640x480) as well as the lower QVGA (320x240) resolution.

Operation	Time (ms)
Transfer of RGB frame to GPU memory	1.385
Transfer of encoded frame to main memory	1.050
Conversion from RGB to YCoCg	1.067
Conversion from YCoCg to RGB	0.402
Contourlet transform decomposition	59.040
Contourlet transform reconstruction	57.102
Run-length encoding of directional subbands	2.424
Run-length decoding of directional subbands	7.008
Contourlet coefficients dropping	0.492

5. CONCLUSIONS

In this chapter, a low complexity algorithm for real-time video encoding based on the contourlet transform and optimized for video conferencing applications and surveillance cameras has been presented and evaluated. The algorithm provides a scalable video compression scheme ideal for video conferencing content as it achieves high quality encoding and increased compression efficiency for static regions of the image, while maintaining low complexity and adaptability to the receivers resources. A video stream can contain various resolutions avoiding the need for reencoding at the source. The receiver can select the desired quality by dropping the components referring to higher quality than needed. Furthermore, the algorithm has the inherent ability to suppress the noise induced by low-quality sensors, without the need of an extra denoising or image enhancement stage, due to the manipulation of the structural characteristics of the video through the rejection of insignificant contourlet transform coefficients. In the case of long recordings for surveillance systems, where higher compression is needed, the visual quality degradation is much more eye-friendly than with other well established video compression methods, as it introduces fuzziness and blurring instead of artificial block artifacts, providing smoother images and facilitating image rectification/recognition procedures. Additionally, due to the relatively small GOPs, the algorithm is more resistant to frame losses that can occur during transmission over IP networks. Another advantageous characteristic of the presented algorithm is that its most computationally intensive parts are calculated on the GPU. The utilization of the usually "dormant" GPU computational power lets the CPU to be utilized for other tasks, further enhancing the multitasking capacity of the system and enabling the users to take full advantage of their computational capabilities. The experimental evaluation of the presented algorithm provided promising results. Nevertheless, in order to compete for compression efficiency with state of the art video compression algorithms, a highly efficient entropy encoding scheme has to be incorporated to the algorithm. Modern entropy encoding methods tend to be complex and computationally intensive. As a result, the optimal trade-off between compression rates and complexity has to be decided in order to retain the low complexity and real time characteristics of our algorithm.

ACKNOWLEDGEMENTS

The authors would like to thank Pavlos Mavridis for his fruitful advice concerning the YCoCg colour space.

REFERENCES

1. S. Katsigiannis, G. Papaioannou, D. Maroulis, 2012 A contourlet transform based algorithm for real-time video encoding Proceedings of SPIE 8437 843704

2. M. N. , M. Vetterli, 2005 The contourlet transform: an efficient directional multiresolution image representation IEEE Transactions on Image Processing 14 12 2091 2106

3. Z. Liu, 2008 Minimum Distance Texture Classification of SAR Images in Contourlet Domain Proceedings of the 2008 International Conference on Computer Science and Software Engineering, CSSE

4. S. Katsigiannis, E. Keramidas, D. Maroulis, 2010 A Contourlet Transform Feature Extraction Scheme for Ultrasound Thyroid Texture Classification Engineering Intelligent Systems 18 3/4

5. Z. Liu, H. Xu, 2010 Image denoising using Contourlet and two-dimensional Principle Component Analysis Proceedings of 2010 International Conference on Image Analysis and Signal Processing, IASP 9-11 April 2010 Xi'an, China 309 313

6. P. J. Burt, E. H. Adelson, 1983 The Laplacian Pyramid as a compact image code IEEE Transactions on Communications 31 4 532 540

7. R. H. Bamberger, M. J. T. Smith, 1992 A filter bank for the directional decomposition of images: Theory and design IEEE Transactions on Signal Processing 40 4 882 893

8. J. M. Shapiro, 1993 Embedded image coding using zerotrees of wavelet coefficients IEEE Transactions on Signal Processing 41 12 3445 3462

9. M. Vetterli, 1984 Multidimensional subband coding: Some theory and algorithms Signal Processing 1984 6 2 97 112

10. Z. Yifan, X. Liangzheng, 2008 Contourlet-based feature extraction on texture images Proceedings of the 2008 International Conference on Computer Science and Software Engineering, CSSE 221 224

11. A. Cohen, I. Daubechies, J. C. Feauveau, 1992 Biorthogonal bases of compactly supported wavelets Communications on Pure and Applied Mathematics 45 5 485 560

12. S. Katsigiannis, 2011 Acceleration of the Contourlet Transform M.Sc. thesis Athens University of Economics and Business

13. P. Mavridis, G. Papaioannou, 2013 The Compact YCoCg Frame Buffer GPU Pro 4 CRC Press

14. H. Malvar, G. Sullivan, 2003 YCoCg-R: A Color Space with RGB Reversibility and Low Dynamic Range Joint Video Team (JVT) of ISO/IEC MPEG & ITU-T VCEG Document No. JVTI014r3

15. D. Van Rijsselbergen, 2005 YCoCg(-R) Color space conversion on the GPU 6th FirW PhD Symposium Faculty of Engineering, Ghent University paper no. 102

16. Center for Image Processing Research (CIPR), Rensselaer Polytechnic Institute http://www.cipr.rpi.edu/resource/stills/index.html accessed 1 July 2012

17. Object and Concept Recognition for Content-Based Image Retrieval Groundtruth Database University of Washington http://www.cs.washington.edu/research/imagedatabase/groundtruth/ accessed 1 July 2012

18. International Telecommunication Union 2011 Recommendation BT.601-7 (03/11) Studio encoding parameters of digital television for standard 4:3 and wide screen 16:9 aspect ratios

CHAPTER 9

Progress of Doppler Ultrasound System Design and Architecture

Baba Tatsuro[1]

[1] Toshiba Medical Systems Corporation, Japan

1. INTRODUCTION

Evolution of electronic technology and semiconductor technology in recent years can realize a high-speed and high-quality signal-processing with low cost, low size, and low power consumption. Various signal-processing devices were born, and their performances are continuing developing. This article introduces the technical innovations and the effects of digital signal-processing in accordance with generations of the Doppler ultrasound system architecture. The diagnostic image of the carotid artery by a Doppler ultrasound system is shown in Fig. 1. The upside image is a tomogram called color flow mapping (CFM). A Doppler range gate is set up in the center of the blood vessel in the CFM. Bloodflow information on this position is displayed as the spectrum

Doppler image in the downside. The horizontal axis is time, and the vertical axis is the flow velocity corresponding to Doppler shift frequency, and it expresses the time-change of velocity distribution of the bloodflow. The embedded technology of CFM and spectrum Doppler began from the composition of analog signal-processing and primitive logical operation elements, and resulted to accumulator devices, PAL, various memories, and changed to FPGA, CPLD, ASIC, DSP, and CPU/GPU [1].

2. PROGRESS OF DOPPLER SIGNAL-PROCESSING ARCHITECTURE

2.1. The 1st generation architecture (Fixed-point processing)

Doppler signal-processing has developed selecting the most suitable realization method in all generations. Architecture of the 1980s is shown in Fig. 2. Analog signal-processing (dark-orange block in Fig. 2) occupied most in this architecture. Henceforth, this is called the 1st generation architecture. Since only fast Fourier transform (FFT) was the digital signal-processing, analog-digital converter (ADC) was arranged before FFT. In those days, the conversion speed of ADC was hundreds kHz in 12-16 bits. Since a complex butterfly-operation was required, FFT processing was realized by accumulators (TRW: 1010J) in the first stage. After a while, a fixed point DSP (Toshiba: DSP-T9508) was used from the second half of the 1980s.

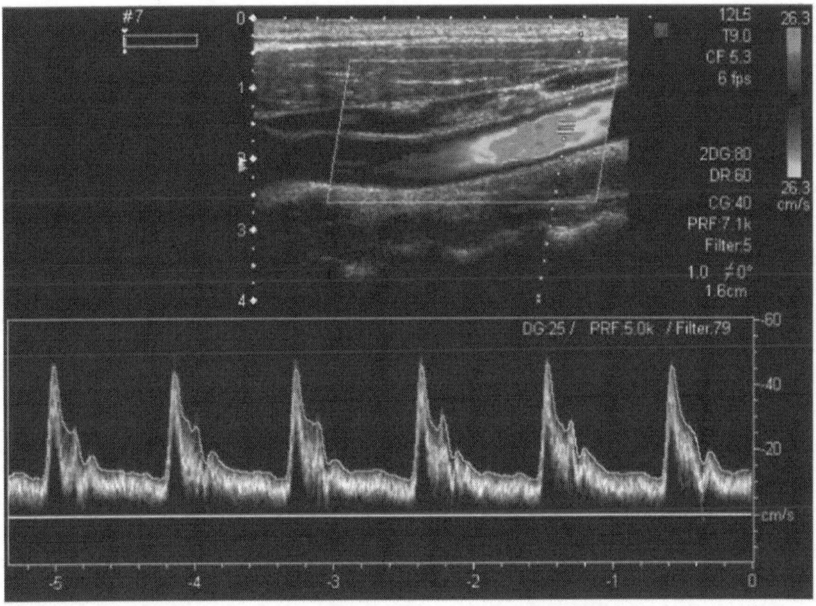

Figure 1. Diagnostic image of Doppler ultrasound system

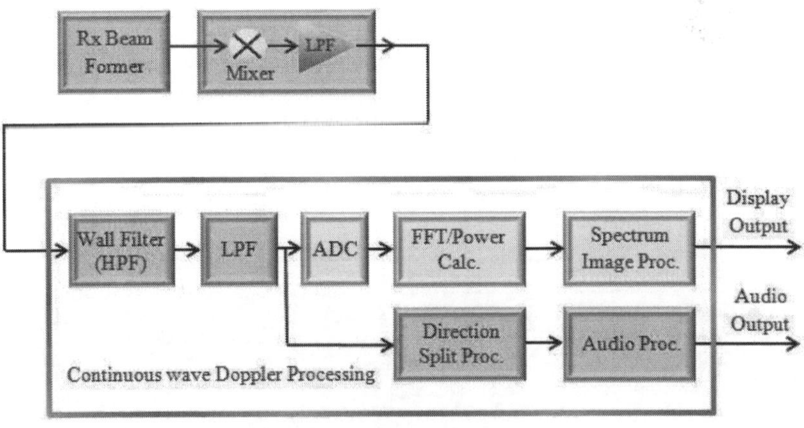

Figure 2. The 1st generation architecture

2.2. The 2nd generation architecture (Floating point DSP and ASIC)

The development of full digital system started early in the 1990s. The early digital architecture is shown in Fig. 3. Henceforth, this is called the 2nd generation architecture. An analog low-pass filter (LPF) was arranged after an analog high-pass filter (HPF) in the conventional analog signal-processing. Since it was difficult to realize a high-speed digital HPF (wall filter in Section 5) with low-cutoff, HPF and LPF were replaced. Furthermore, the LPF with high sampling frequency was divided into two subcomponents. The first stage LPF was realized by FPGAs (Altera: FPGA), and the next stage LPF and HPF were realized by floating point DSPs (NEC: μPD77240A), respectively. Furthermore, FFT processing was realized by an ASIC (Toshiba: ASIC). However, as for Doppler audio processing (direction separation of complex signal, etc.), the conventional analog-circuit was used in consideration of cost-performance. Therefore, the digital filter output was converted into analog signal again by digital-analog convertoer (DAC), and was inputted into analog-circuit.

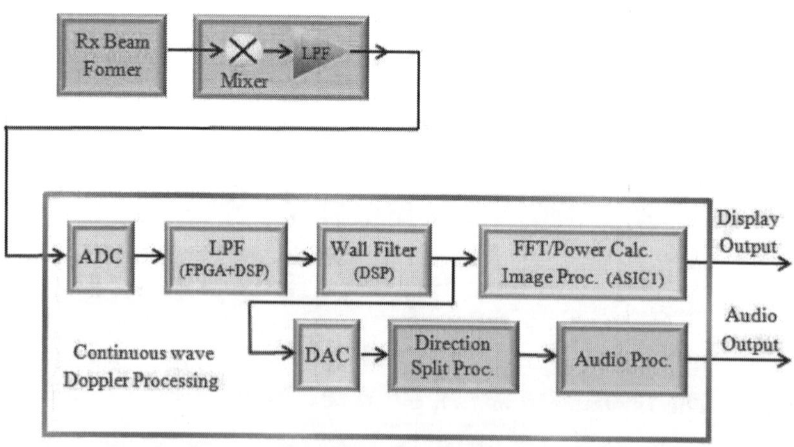

Figure 3. The 2nd generation architecture

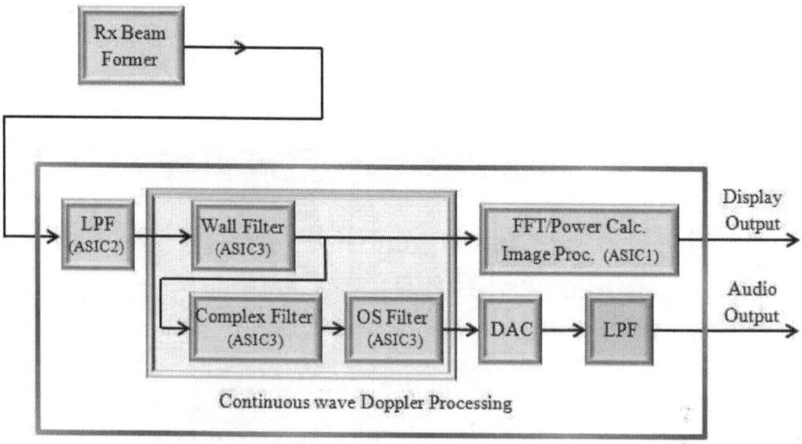

Figure 4. The 3rd generation architecture

2.3. The 3rd generation architecturc (Dynamic-range expansion by ASIC)

In the second half of the 1990s, in order to merge CFM and spectrum Doppler, the development which reduces the size and cost of these systems started. Henceforth, this is called the 3rd generation architecture. In order to compare generations, only a spectrum Doppler portion is shown in Fig. 4. In the 3rd generation, since the system clock went up sharply, the floating point device was hard to use. CFM and spectrum Doppler were unified, and they were realized by five kinds of ASIC (Toshiba: fixed point ASIC). In this architecture, I adopted the newly developed digital complex IIR filter for the direction separation processing without using analog phase-shifters with heavy manual adjustments [2, 3]. Furthermore, an oversampling filter and high-speed DAC (Analog Device: DAC) were used for the Doppler audio processing. This reduced the analog-circuit, such as high-order switched capacitor filters (SCF). The scale of these large-scale ASIC reached more than twice of typical CPU (Intel: Pentium processor) respectively, but the total cost of spectrum Doppler declined in 1/3. Furthermore we were able to get the wide dynamic-range signal-processing which was difficult in analog processing. As a result, the sensitivity of bloodflow detection had improved and the diagnostic targets also spread to abdomen, surface blood vessels, and limbs etc.

2.4. The 4th generation architecture (Reduction of circuits by large scale DSP)

In the first stage of the 2000s I realized whole Doppler signal-processing using only one floating point DSP (TI: DSP TMS320C6701). The signal-processing block inside DSP is shown in Fig. 5. Henceforth this is called the 4th generation architecture. Since the clock frequency went up tens times compared with the 2nd generation floating point DSP, throughput improved sharply. Moreover, changes of the ultrasound system architecture contributed to downsizing. The interrupt cycle to DSP was changed into display frequency (Vsync: 50-75 Hz) from ultrasonic pulse repetition frequency (PRF: 1-50 kHz). Although real-time performance was spoiled a little by forming packet processing, drastic reduction of circuit scales was realized.

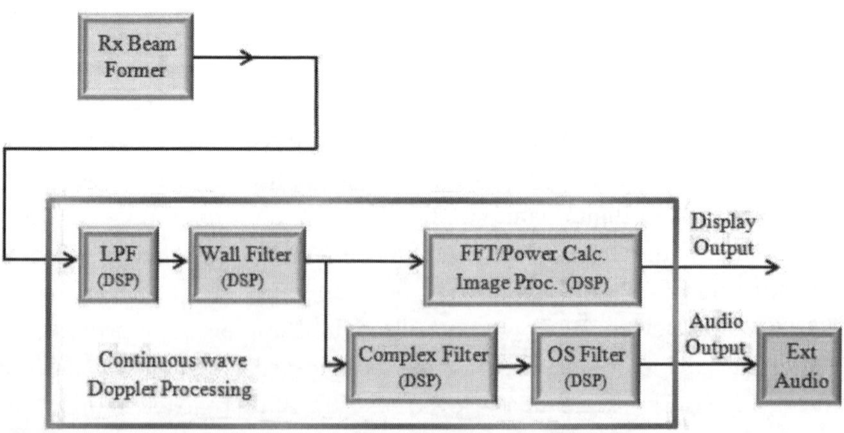

Figure 5. The 4th generation architecture

2.5. The future architecture (Real-time analysis)

Evolution of signal-processing had been influenced by the realization methods, like as from analog to digital, or from hardware to software. The size and cost of Doppler signal-processing were reduced day by day, and its performance (sensitivity or the response etc.) was also improved. It will be possible to gain more huge calculation power for signal-processing from now on. It will be also an important theme to investigate for what this power should be used. Now I am trying to apply this ability to automatic measurement and automatic diagnosis as an intellectual signal-processing [4-6]. Moreover, as another possibility, mounting the Doppler signal-processing on a Windows program is also tried. But problems, such as stability or response time, remain. In order to realize the real-time system which completes the processing within time, I think that the architecture based on DSP will remain.

3. CONSIDERATIONS OF DIGITAL TECHNOLOGY

As introduced in Section 2, the spectrum Doppler signal-processing architecture had changed from analog to digital in the first half of the 1990s. In this section, comparison of analog technology and digital technology and the innovations which digital technology brought about are introduced.

3.1. Comparison of analog processing and digital processing

As compared with analog processing, merits of digital processing are shown in Table 1. Digital processing realized quality improvement (reduction of variations), performance improvement (wide dynamic-range), and size reduction. Moreover, development efficiency was improved by the separation of analog power supply and digital power supply, and by the reduction of the noise in analog systems. Also the cost of digital-circuit had been improved in the 1990s more easily than analog-circuit.

Table 1. Comparison of analog processing and digital processing

Items	Analog Processing	Digital Processing
Variation caused by electrical parts	Big. (Expensive adjustment for analog-circuit)	Small.
Tolerance to noise	Noise countermeasure must be done to every sub-block. Limitation of dynamic-range	Only ADC needs adjustment. Wide dynamic-range is realizable.
Kinds of power	At least 3 kinds The isolation between digital Power supply and analog power supply is needed.	Only the digital power except for ADC and DAC
Cost performance	Before the 1990s, analog processing had high C/P ,but recently, it is low C/P compared with the digital processing.	Cost, size and power consumption are improving substantially every year.
Throughput	Real time performance is good.	Digital processing realizes the high-speed and complex processing.

3.2. Time-spatial resolutions and S/N ratio

By development of digital technology, sampling frequency and pixel size are increasing in a digital camera and a digital audio field every year. When sampling frequency and pixel size increase, finer sampling becomes possible in time and space. The spatial-resolution and the time-resolution have improved recently, and a high-definition image and a high-fidelity audio can be enjoyed now. Moreover, the product performance that exceeds human vision and hearing is also improved. However, from the viewpoint of manufacturing cost, if the target performance to demand is filled, the present performance level may be enough. But the products which exceed this performance actually appeared one after another in the market. I consider this reason as follows. The present product level does not fill the dynamic-range of human vision and hearing. Since human sensitivity perceives physical quantity by logarithm, according to the surrounding environment, a wide dynamic-range is required. That is, I think that the commercial products of digital camera or digital audio have not reached the demand dynamic-range of luminosity or sound pressure yet [7-9].

Aside from improvement in spatial resolution or time resolution, another merit of digitization is a high S/N ratio. A digital signal is sampled in a spacial axis and/or a time-axis. An ensemble mean processing can remove the noise which adjoins in space or time, so it can extract a low-frequency component with sufficient accuracy. An ensemble mean model, a signal level, a noise level, and expansion of S/N ratio are shown in Fig. 6. When the ensemble number is set to *N*, a signal increases to *N* times and a noise increases to $\sqrt{N}$ times so an S/N ratio is expanded to $\sqrt{N}$ times [10, 11]. Bandwidth restriction filters (HPF, LPF, BPF etc.) has the same effect of the ensemble mean processing. In digital processing designs, we should be cautious of an internal dynamic-range and an output noise level. In analog system, since the dynamic-range was narrow, a wide dynamic-range signal was not processed faithfully. The artifact caused by saturation or quantization occurred in the intermediate processes, so the sufficient sensitivity of bloodflow detection was not obtained. After the 3rd generation architecture, as the wide dynamic-range signal-processing was realized, the big improvement in bloodflow diagnosis was brought about.

3.3. Hardware reduction by over-sampling

The design concept of the compact disk (CD) which started in the 1980s was that the stereo digital signal (44 kHz sampling) was changed into the stereo analog signals (maintaining 20 kHz bandwidth). For this purpose, the steep analog filter (at least 7th-order) which rejects harmonics after AD conversion output (22 kHz) was required. Several years afterward, 4-times over-sampling system (about 170 kHz) appeared. It was realized by a digital moving average filter (sampling-frequency: about 170 kHz, cutoff frequency: 20 kHz, cutoff property: loose), a 4 times DA conversion, and a simple analog filter (cutoff frequency: 20 kHz, cutoff property: about 2nd-order). Since this system had many merits (reduction of cost and size, improvement of S/N ratio etc.), it became mainstream [12].

The third generation architecture of Doppler ultrasound system was designed based on this over-sampling concept. In Doppler audio processing, unlike CD, a sampling frequency changes widely (1 kHz to 50 kHz). Therefore, the cutoff frequency of digital filter before DA conversion had to be variable. The change range of sampling frequency is the same as not only Doppler audio processing but also digital filters of the 2nd generation. The effect of the over-sampling processing is shown in Fig. 7. Fig. 7(a) shows the sampling characteristic of the sampling frequency *fs*, and harmonic (a side lobe of -14 dB) is mixed because of a simple over-sampling (hold characteristic). In order to remove these harmonics and to keep the required dynamic-range in required bandwidth, the filter with suitable bandwidth property (broken line) is required. In the case of Fig. 7(b), since *fs/BW* is small, a high-order filter is required. But in the case of Fig. 7(c), since *fs/BW* is large, a low-order filter is also fully realizable. The sampling frequency of high-speed digital devices is going up now. Since *fs/BW* is expanded, both downsizing and high-performance are realized simultaneously [13].

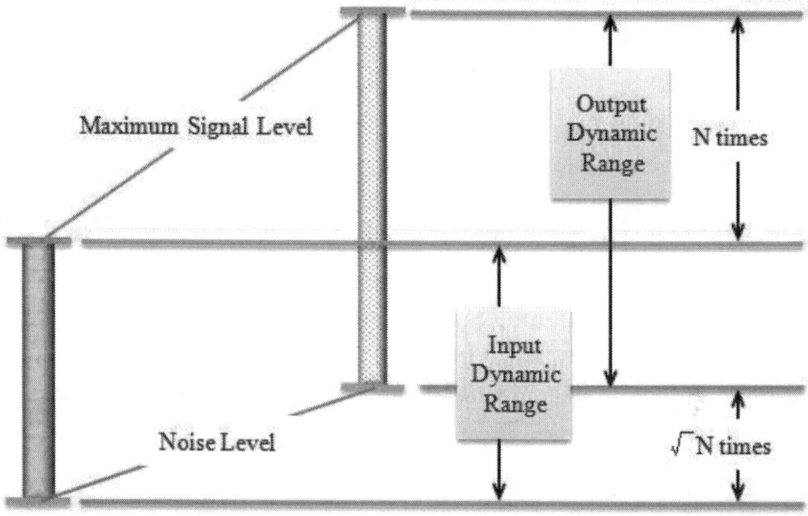

Figure 6. Principal of ensemble mean processing and S/N ratio expansion

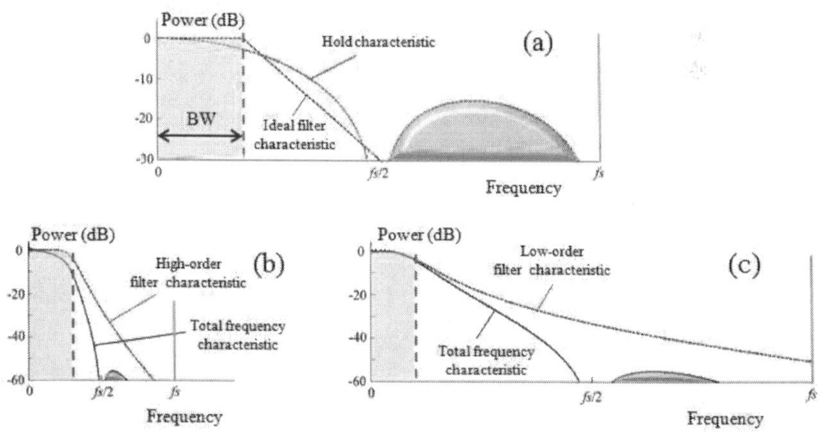

Figure 7. Over-sampling processing (a) Over-sampling model (b) Small *fs/BW* case (c) Large *fs/BW* case

4. CASCADE DIGITAL FILTER DESIGN

The design tools for various digital filters, such as LPFs, HPFs, and BPFs, were popular and spread. However, in order to realize large-scale signal-processing combining various filters which have different sampling frequency, there are many points that should be taken into consideration, such as an aliasing and artifacts. In this section, two design examples of the cascade digital filter are introduced. One is a cascade BPF for continuous wave Doppler processing that has the down-sampling processing exceeding 10^{-4} or less. Another is a cascade LPF for Doppler audio processing that has the over-sampling processing exceeding 10^{3} or more.

4.1. Continuous wave Doppler signal-processing

In the Doppler signal-processing, an HPF is effective to extract weak bloodflow. There is a clutter signal which mixes unnecessary reflective ingredients from a blood vessel wall etc. By removing it at the entrance of frequency-analysis, the dynamic-range of FFT processing and audio processing can be held down, and also the signal-processing load can be reduced. High-order HPF array (bandwidth: several kHz, cutoff-frequency: several hundred to several thousand Hz) had been used in the analog system. For digitizing this, the 2nd generation architecture was developed. The filter array for continuous-wave Doppler signal-processing is shown in Fig. 8.

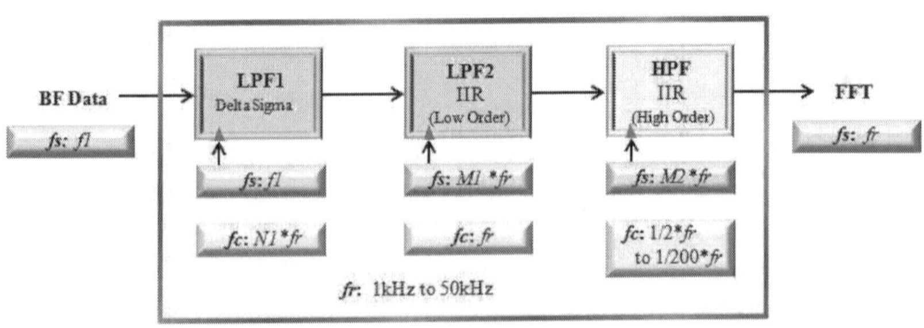

Figure 8. Digital filter design of continuous-wave Doppler signal-processing

In the conventional system, after an analog HPF, a bandwidth restriction was applied by the anti-alias LPF before ADC. To realize the digital HPF which had high-order and high sampling frequency was difficult in those days. For example, when a high-order IIR filter (several MHz and 2ch processing) was assumed, hundreds Mflps performance was required for calculation. Then, LPF was arranged before HPF in digitization. A high-speed and low-cutoff LPF was required. For example, the relative cutoff 1/1000 (that means several kHz bandwidth restrictions with tens of MHz sampling) was required. In order to prevent expansion of the tap-length (number of delay-registers) and bit-length of internal registers (inner dynamic-range), I chose the system which divided LPF into two steps and applied down-sampling. Since the output after quadrature detection was high-speed of tens MHz (*f1* Hz in Fig. 8), in the front part of LPF the delta-sigma LPF that had cutoff frequency *N1*fr* was adopted, and it was realized by FPGA.

The latter part of LPF had re-sample frequency *M1*fr*, and it was realized by the single-precision floating point DSP (low-order LPF, cutoff-frequency: *fr*). An HPF was arranged after these two steps of LPFs. The HPF carried out scaling-processing to the LPF output by re-sampling frequency *M2*fr*. High-order and wide-range cutoff HPF processing (cutoff-frequency: *fr*/2 to *fr*/200) was realized by the double-precision floating point DSP. By this architecture, the required bandwidth restrictions and dynamic-range could be realized even in the continuous-wave Doppler processing which had heavy mixing of clatter artifacts. The frequency characteristic of the cascade digital filter (Fig. 8) is shown in Fig. 9. Actually a chirp waveform (0 to 40 kHz) was inputted into the cascade digital filter and its performance was checked by the spectrum Doppler image of the trial product. We can check the loose bandwidth restrictions near ±6 kHz (*fr*) in this figure.

4.2. Doppler audio signal-processing

After quadrature detection, the Doppler audio system divides IQ signal into a forward component and a reverse component with a direction separation filter, and outputs them to the left and right stereo speakers.

In the case of a pulse wave Doppler, since about 4 kHz sampling frequency is interlocked with *fr* (pulse repetition frequency), it is hard to hear bloodflow signal as it is (by the mixing of harmonics). In order to remove harmonics, it is required to realize the steep filter which has cutoff of *fr*/2 in consideration of audio bandwidth (20 Hz to 20 kHz), and to reject unnecessary harmonics. The conventional analog filter architecture is shown in Fig. 10. In the Fig. 10, LPF1 had the steep cutoff characteristic by SCF after direction separation processing. And LPF2 removed the harmonics generated in LPF1 (SCF noise). The S/N ratio of SCF was 50 dB or less in audio-range, and sound quality was quite bad compared with the present system.

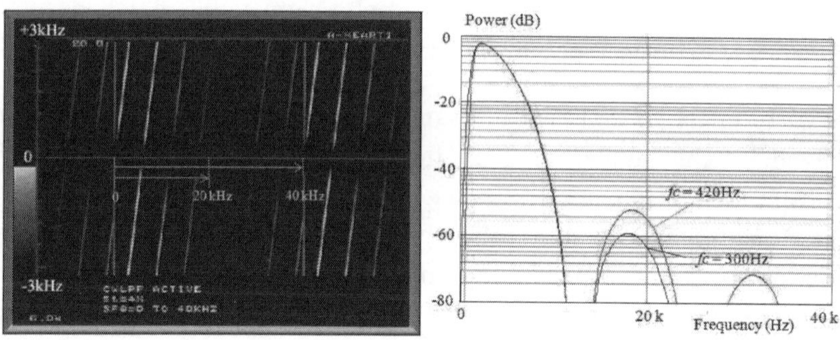

Figure 9. Frequency characteristics of continuous-wave Doppler filter of the 2nd generation architecture

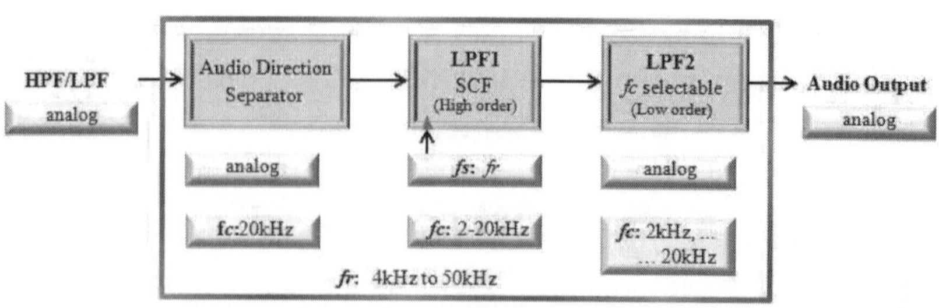

Figure 10. Analog Doppler audio system of the 2nd generation architecture

The Doppler audio filter of the 3rd generation is shown in Fig. 11. The signals of direction separation (complex BPF output) were oversampled by LPF1 whose sampling frequency (*M3*fs*) was hundreds times larger than *fr*. And LPF1 had a loose bandwidth restriction by a moving average. In the following stage LPF2, scaling was applied by *f2* (same as the ADC clock frequency). LPF2 had the bandwidth restriction by the IIR filter (cutoff frequency: *fr*/2). After AD conversion, in order to remove harmonics, loose bandwidth restriction was again applied by LPF3. This cascade filter processing could realize a high-quality Doppler audio (S/N ratio: more than 90 dB). Moreover, since the oversampling frequency of LPF1 and the conversion frequency of ADC were set up more highly, the simple filter (lower-order) was used and drastic hardware reduction of LPF2 and LPF3 were realized [14].

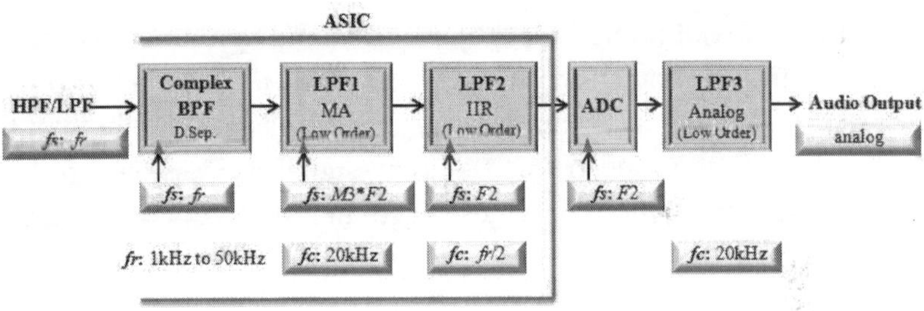

Figure 11. Digital Doppler audio system of the 3rd generation architecture

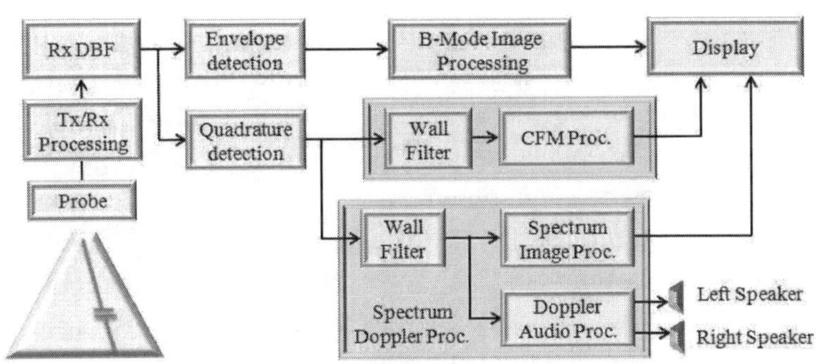

Figure 12. Wall filter arrangements of Doppler ultrasound system

5. HIGH-PRECISION DIGITAL FILTER DESIGN

Doppler ultrasound diagnostic method spread to many diagnostic fields, such as cardiac and abdomen. On the other hand, the improvement of bloodflow detection (sensitivity and velocity-range) had been desired for a long time. For this purpose, since it was required to separate weak bloodflow signal from high-power artifacts, like a blood vessel wall, the steep HPF (called a wall filter) had been arranged before frequency-analysis. With digitization, I investigated a new wall filter designs.

5.1. Purpose of wall filter

The locations of wall filters in Doppler ultrasound system are shown in Fig. 12. Quadrature detection outputs (IQ signal) are divided into CFM processing and the spectrum Doppler processing. In both CFM processing and the spectrum Doppler signal-processing, to save dynamic-range, wall filters are arranged before frequency analyses respectively [15].

Beam scanning methods and the display modes corresponding to them are shown in Fig. 13. Fig. 13(a) shows the scanning method of a B-mode echo image (tomogram), and it scans a beam from the right to the left. Fig. 13(b) shows the scanning method of spectrum Doppler, and it scans the same beam in a tomogram continuously. Fig. 13(c) shows the scanning method of CFM, and it scans a beam from the right to the left like Fig. 13(a), but the same beam is scanned twice or more. The sampling methods of Fig. 13(b) and Fig. 13(c) are shown in Fig. 14(a) and Fig. 14(b). Beam data is sampled by *fr*, and includes the information on the depth direction. As shown in Fig. 14(a), since the spectrum Doppler has a long time series signal, a detailed frequency analysis can be realized. On the other hand, as shown in Fig. 14(b), CFM has plurality data series (hundreds points) on the depth direction. Because CFM processing consists of a finite wall filter and a complex autocorrelation processing, the analysis data of CFM is same as sampling beam number

(5 in the case of Fig. 14(b)), and is very small compared with that of the spectrum Doppler.

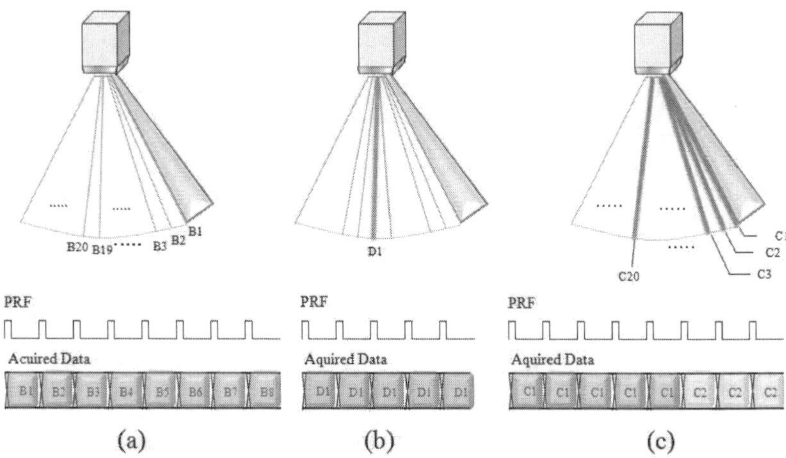

Figure 13. Ultrasound beam scan mode (a) B-mode (b) Spectrum Doppler (c) CFM

The clutter is the high-power and low-frequency component at the quadrature detection output. Compared with a bloodflow signal, it has high-power low-frequency component in abdomen (about 20 dB) and heart (more than 40-60 dB). The quadrature detection outputs and wall filter outputs collected in the heart (left ventricle outflow) are shown in Fig. 15. The horizontal axes of Fig. 15(a) and Fig. 15(b) are time, and vertical axes are amplitude. The horizontal axis of Fig. 15(c) is frequency and a vertical axis is power. The wall filter has the 4th-order Butterworth characteristic with 200 Hz cutoff frequency. The power spectra of Fig. 15(c) show that a big clatter component (20dB bigger) is removed. The wall filter is required high-order (steep) and low-cutoff characteristic [16].

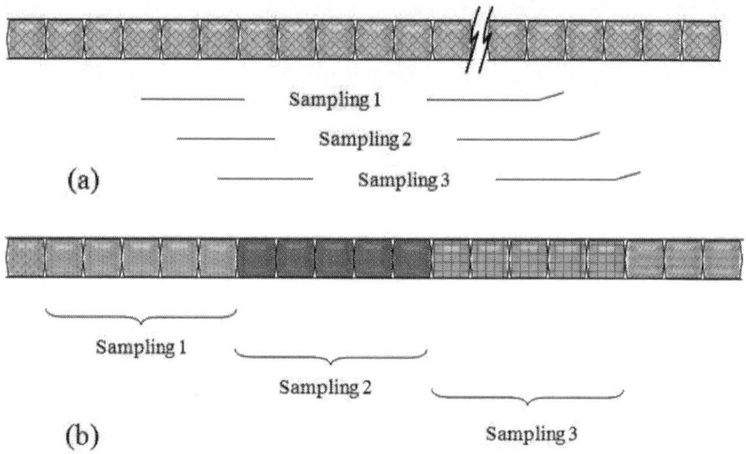

Figure 14. Sampling methods (a) Spectrum Doppler (b) CFM

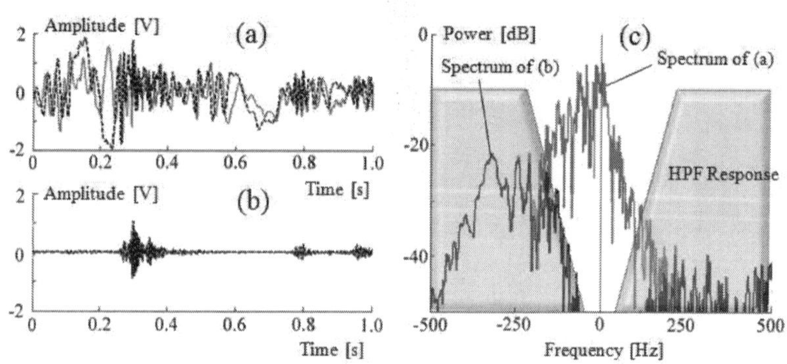

Figure 15. Effect of wall filter (a) Input signals of wall filter (b) Output signals of wall filter (c) Spectra of (a) and (b)

5.2. Wall filter of CFM

The high-order analog filter had been used for the wall filter of spectrum Doppler, and a finite digital filter had been used for the wall filter of CFM for a long time. The step responses of infinite impulse response (IIR) Butterworth filter in the case of changing cutoff and order are shown in Fig. 16. Fig. 16(a) shows the responses when order is changed at the relative cutoff frequency 1/16 (normalized by sampling frequency), and Fig. 16(b) shows the responses at the relative cutoff frequency 1/128. A transient response becomes long at the time of low-cutoff and high-order (steep). Relation between cutoff and order when the transient response is set to -20 dB (10% of step input amplitude) is shown in Fig. 16(c). Since performance of the wall filter with finite input is insufficient, the technology for reducing a transient response was required. The wall filter systems of CFM are shown in Fig. 17. In the finite impulse response (FIR) system of Fig. 17(a), if the number of delay registers N is small, sufficient performance cannot be obtained. So the IIR system of Fig. 17(b) became main-stream. However, unlike the wall filter of spectrum Doppler, the transient response of IIR filter has a serious influence to frequency-analysis. In order to solve this problem the adaptive filter which is consistuted by a time-variant FIR filter shown in Fig. 17(c) appeared recently.

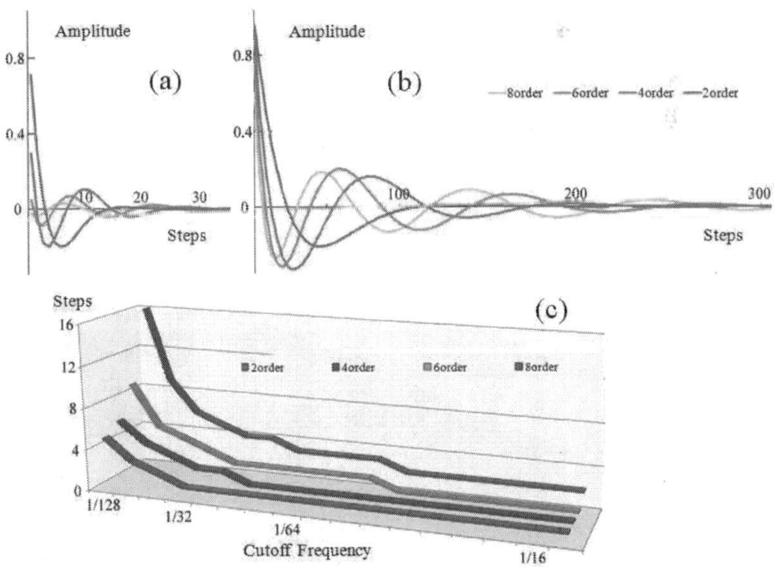

Figure 16. Step responses of HPF (a) Responses @ relative cutoff: 1/16 (b) Responses @ relative cutoff: 1/128 (c) Cutoff and transient response

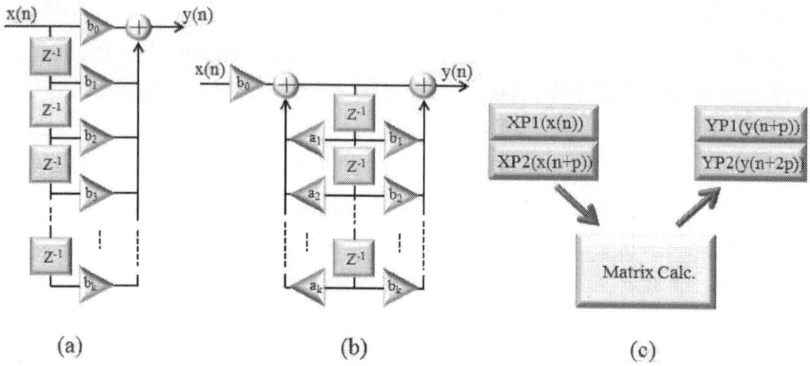

Figure 17. Wall filter systems of CFM (a) FIR filter (b) Biquad filter (c) Time valiant FIR filter

Since there is no feedback in the FIR system of Fig. 17(a), saturation does not occur easily, and the internal dynamic-range can be made small. However, since it is necessary to increase the number of taps in order to obtain low-cutoff, it is disadvantageous in a response and size. Biquad filter system of Fig. 17(b) has good response. However, since big internal dynamic-range is required at the time of low-cutoff, there are problems of a quantizing noise or a transient response. Compared with these, although time delay is given as a packet unit, a time-variant FIR filter system shown in Fig. 17(c) has many advantages. The filter response is calculated from packet data based on matrices. By progress of signal-processing device in recent years, the development of an adaptive filter based on the time-variant FIR system became also easy. A time-variant FIR filter with input x(n), output y(n), and the state variable *v(n)* shown in Fig. 17(c) consists of the state equation and output equation in equation (1).

$$\begin{aligned} v(n+1) &= F * v(n) + q * x0(n) \\ y(n) &= g^T * v(n) + d * x(n) \end{aligned} \tag{1}$$

The coefficients of an IIR filter (fig. 17(b)) are transposed into the matrix *F*, *q*, *g* and *d* in equation (2). The signal-processing equivalent to an IIR system can be realized by a time-variant FIR system.

$$F=\begin{bmatrix} 0 & 1 & 0 & \cdots & 0 \\ 0 & 0 & 1 & \cdots & 0 \\ & & \cdots & & \\ 0 & 0 & 0 & \cdots & 1 \\ -a_K & -a_{K-1} & -a_{K-2} & \cdots & -a_1 \end{bmatrix} \qquad q=\begin{bmatrix} 0 \\ 0 \\ \cdots \\ 0 \\ 1 \end{bmatrix}$$

$$g=\begin{bmatrix} b_K - b_0 * a_n \\ b_{K-1} - b_0 * a_{K-1} \\ \cdots \\ b_1 - b_0 * a_1 \end{bmatrix} \qquad d = b_0 \tag{2}$$

Thus, since the time-variant FIR system shown in Fig. 17(c) can solve both the problem of the internal dynamic-range and a transient response, it will develop as a new wall filter of CFM from now on.

5.3. Wall filter of spectrum Doppler

The step responses of 8th-order Butterworth filter (4 cascade biquads, relative cutoff 1/128) are shown in Fig. 18. Fig. 18(a) shows the step input *x(n)* and the output *y(n)*. Since it is HPF, its output approaches DC with a damped oscillation. The responses of the inner registers for each stage (*Z1(n)*, *Z3(n)*, *Z5(n)* and *Z7(n)* in Fig. 20(b)) are shown in Fig. 18(b). Although *Z3(n)*, *Z5(n)* and *Z7(n)* are converged on DC with about tens times amplitude of an input, *Z1(n)* holds about 400 times amplitude of an input. Thus, when the HPF prevents saturation or keeps internal accuracy, wide dynamic-range of internal registers is required. The relation between cutoff and the dynamic-range (the bit-length of inner register) is shown in Fig. 19. In order to realize a high-precision digital filter, accuracy of operation registers and filter coefficients is important. I checked the minimum bit-length that was not influenced by quantizing noise. The responses of the fixed point 8th-order Butterworth filters were simulated. If quantizing noise is mixed, the unstable oscillation such as a limit cycle etc. will occur. I changed cutoff frequency and measured the limit of stability. As a result, in order to realize low cutoff, it turned out that sufficient mantissa-length of operation-registers and sufficient multiplication-coefficient length of multipliers were required. In fact, since cutoff frequency became about 1/200 in the spectrum Doppler processing, a huge internal dynamic-range (about 200dB) was required.

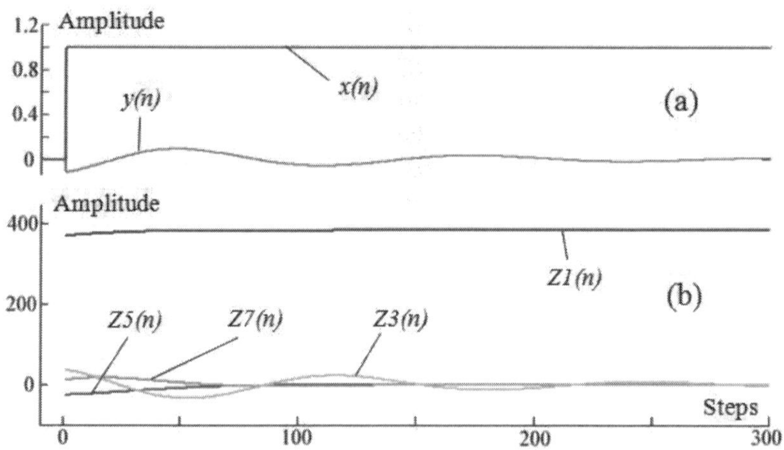

Figure 18. Step response of HPF and transit response (a) Input *x(n)* and output *y(n)* (b) Responses of internal resisters

The signal-processing architecture which reduces the internal dynamic-range of a digital filter was developed in the 2nd generation architecture. Simultaneously, the algorithm which reduces calculation in a real-time system was also investigated. The systems to realize 8thorder digital filter are shown in Fig. 20. Fig. 20(a) shows the loop system which makes the internal dynamic-range small with four delay-registers. Fig. 20(b) shows the loop biquad filter system with two delay-registers. The upper Z^{-1} corresponds to *Z1(n)*, *Z3(n)*, *Z5(n)* and *Z7(n)* of Fig. 18(b). Fig. 20(c) shows the system with eight delay-registers in series. While the calculation cycle becomes small, the dynamic-range of internal registers becomes large.

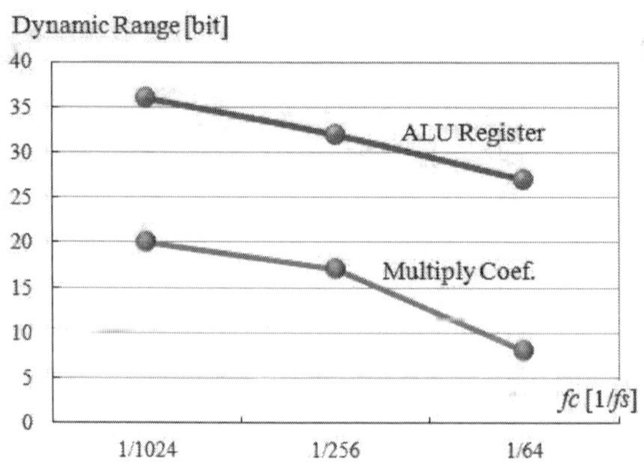

Figure 19. Dynamic-range (bit-length) and cutoff frequency of HPF

Fig. 21 shows the evaluation of above systems. As benchmark condition, NECμPD77240 was used for floating point DSP. The 8th-order Butterworth HPF was chosen as benchmark processing, and its relative cutoff was set to 1/256. In the system 3 (equivalent to Fig. 20(c)), although both the number of double-precision registers and the operation cycles were small, internal bit-length became large. Since inner bit-length exceeded above 50 bits, even the double floating point arithmetic (mantissa-length: 48 bits) run short of accuracy, and was difficult to realize. In the system 1 (equivalent to Fig. 20(a)), many operation registers were required although internal bit-length was small. Since low-speed external memory access was required, its operation cycle increased. In the system 2 (equivalent to Fig. 20(b)), internal bit-length did not exceed the range of double-precision floating-point arithmetic, and an operation cycle was comparatively small. As mentioned above, the system 2 was judged the best system for mounting [17].

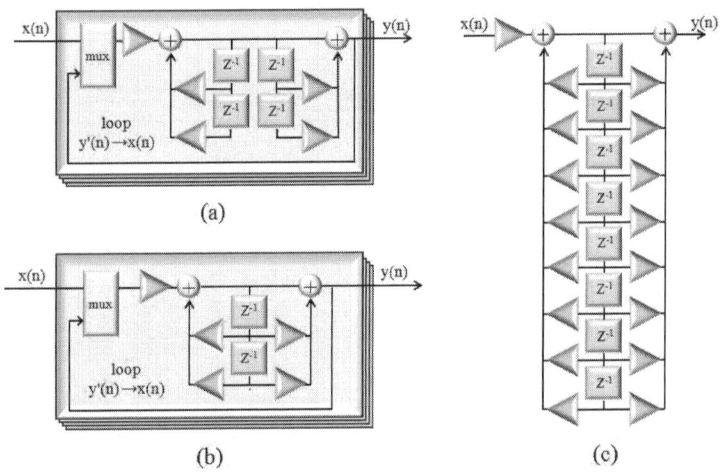

Figure 20. Wall filter systems of spectrum Doppler (a) Sysytem1: IIR+FIR loop system (b) System2: Biquad loop system (c) System3: Direct IIR system

6. WIDE DYNAMIC-RANGE SYSTEM DESIGN

The dynamic-range of conventional system was insufficient for some clinical applications. In recent years, diagnostic ultrasound system was improved through the use of high-frequency electronics and integrated circuits. A new diagnostic method became effective by the higher dynamic-range system. However, the higher dynamic-range system also means more complicated gain control. It is possible to optimize the gain automatically through the use of ultrasound system parameters. This technology reduces the size of hardware and reduces the gain control range substantially.

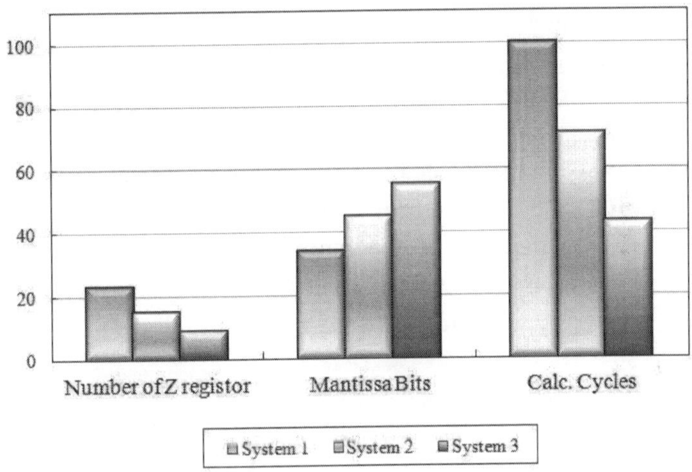

Figure 21. Benchmark based on floating point DSP (μPD77240)

6.1. Signal-processing of ultrasound system

Conventional ultrasound signal-processing is shown in Fig. 22. The transceiver processor (Tx/Rx Proc.) receives signals from the probe elements and the receive signals are amplified by the preamplifier. Gain compensation is applied to the signal to correct for range-distance attenuation (STC: sensitivity time control) and an analog gain correction for probe characteristics (frequency, sensitivity, etc) is applied. The signal is then sent to an ADC. After AD conversion the digital beam former (DBF) applies a delay pattern to the data to focus it and produce beam data. This data is processed by B-Mode Image Proc. and Doppler Image Proc., then displayed as a tomogram image and/or a spectrum Doppler image in the display processor (Display). The S/N ratio is increased in the DBF as the number of channels (corresponds to transducer elements) to which delay calculations are applied increases [18, 19]. In Doppler signal-processing, quadrature detection (Mixer) is applied to the DBF output and a BPF provides band-limitation and clutter rejection. The result is a base-band Doppler signal. At this time, the S/N ratio is sharply increased because of the band-limitation of the BPF. In the Doppler signal-processing, I apply range-gate integration (RG) across the range direction of the ROI. This also increases the S/N ratio. In the case of continuous wave Doppler the dynamic-range is even larger. And the HPF applied to this data must be more sophisticated. The dynamic-range of the signal leaving the HPF is also much larger than in the pulse wave case, in the order of 100 dB. After FFT the S/N ratio is greatly increased because of the butterfly integration. The dynamic-range of the signal is now very large and considerable gain adjustment and display compression must be done in order to display the data.

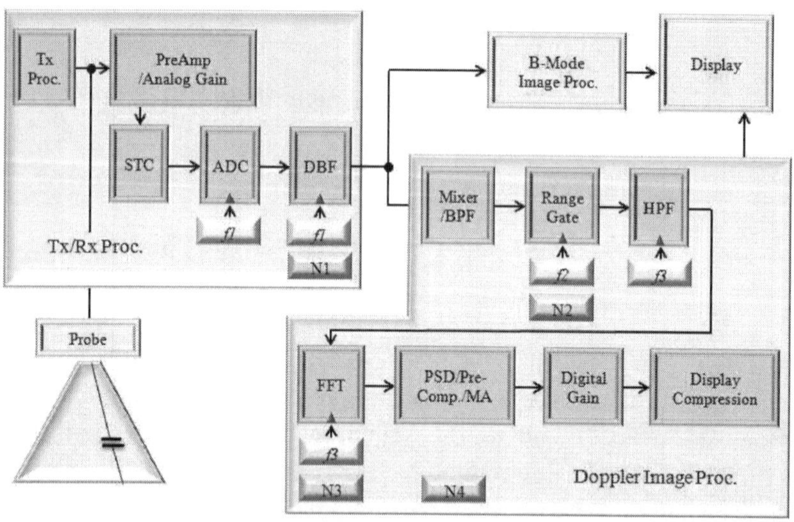

Figure 22. Ultrasound signal-processing system

It is also necessary to take into consideration that the dynamic-range is increased in Doppler signal-processing. In addition, mirror effect and/or quantization artifacts are introduced when performing automatic gain compensation. Although the beam data leaving the DBF has a frequency in the order of about 10-100 MHz, it is re-sampled at about 1-100 kHz. So the input dynamic-range of FFT increased by the band-limitation effect. Moreover the S/N ratio of FFT output is increased in a similar manner to that of ensemble mean processing.

6.2. Purpose of gain adjustment

Gain adjustment corrects for diagnostic target, bloodflow sensitivity, and the difference of user's skill. In addition to this, gain adjustment compensates for variation in other equipment parameters, such as the number of summing channel in DBF, apodization function, bandwidth of the BPF, the integration length of the range-gate (RG), FFT number, window function, and the number of the shift addition of power spectrum according to sweep speed etc. The maximum signal level and a noise level change because of change of these equipment parameters. In order to realize highly sensitive Doppler bloodflow diagnosis without saturation, a system with wide dynamic-range must perform gain compensation according to all these parameters. Table 2 illustrates the rough estimation of the dynamic-range and S/N ratio based on the virtual system [20].

The model of the signal-processing accompanied by expansion of the S/N ratio is shown in Fig. 23(a). The noise level and the maximum signal level of the incoming signal are expanded by signal-processing. But the expansion of the noise level differs from the expansion of the maximum signal level, and the overall S/N ratio is increased. Under optimal gain adjustment (range shown in light green in Fig. 23(a)) that there is no saturation of the maximum signal level and quantizing noise and signal are not mixed in the output. When gain adjustment is unsuitable (range shown in light pink in Fig. 23(a)), mirror effect or quantization artifacts occur on the spectrum image due to saturation or omission. As for signal amplitude, when quantization accuracy is inadequate, quantizing noise mixes. The gain adjustment that takes into consideration when detecting the weak Doppler signal around a system noise, and has sufficient quantizing margin is required. The influence of quantization is shown in Fig. 23(b). It is a spectrum when inputting sinusoidal (0.02**fs*) including white noise. The horizontal axis is time and the vertical axis is frequency normalized by a sampling frequency *fs*. The quantizing level of an input range was changed every 2 seconds with 3, 5, 9, 17. It turns out that the harmonics components (-20 to -30dB) by quantization has occurred near the frequency -0.3**fs*, -0.2**fs*, and +0.25**fs*. The mirror effect is an imaginal image symmetrically generated with a real image on both sides of a baseline. In the analog system, it is mainly caused by the phase error of quadrature detection, or the small gain difference between IQ signals. In a digital system, although these influences do not receive, a mirror effect generates them owing to saturation. As shown in Fig. 23(c), on both sides of 0 Hz, a symmetrical mirror effect occurs in the spectrum image. Fig. 23 (c) is the spectrum image which raised the gain 6 dB at a time every 2 seconds to the sinusoidal input including white noise. In this figure FFT input dynamic-range is 16 bit. The horizontal axis is time and the vertical axis is the frequency that normalized by a sampling frequency fs. The mirror effect component (-0.2**fs*) has occurred by saturation to an original signal component (+0.2**fs*). So in conventional ultrasound design both the mirror artifact and quantization artifact are caused by insufficient system dynamic-range in Doppler signal-processing.

Table 2. Comparison of inner dynamic-range and gain control range

Module	Cause	Effect of D.R. Increment (*1)	Conventional System Fig. 24(a) (dB)	New System Fig. 24(b) (dB)
ADC output	(Analog Gain)	50dB	DR1 @f1	DR1 @f1
DBF	BeamSum effect(*N1* channel)	+50dB	DR2=DR1 + 20log(*N1*)	DR2_opt=DR1 + 20log(√*N1*)
Mixer/BPF	BandLimitation effect (*f1/f2*)	+30dB	DR3=DR1+DR2 + 20log(*f1/f2*)	DR3_opt=DR1+DR2_opt +20log(√(*f1/f2*))
RG	RG Integration effect(*N2* tap)	+40dB	DR4=DR1+DR2+DR3 + 20log(*N2*)	DR4_opt=DR1+DR2_opt +DR3_opt + 20log(√*N2*)
FFT	FFT number and window(*N3* sampling)	+50dB	DR5=DR1+DR2+DR3+DR4 + 20log(*N3*)	DR5_opt=DR1+DR2_opt +DR3_opt +DR4_opt + 20log(√*N3*)
PSD/Pre-Compres.	Power to Amplitude	-	DR5	DR5_opt
MA	Moving Average effect(*N4* average)	+10dB	DR6=DR1+DR2+DR3+DR4 +DR5 + 20log(*N4*)	DR6_opt=DR1+DR2_opt +DR3_opt +DR4_opt+DR5_opt + 20log(√*N4*)

Module	Cause	Effect of D.R. Increment (*1)	Conventional System Fig. 24(a) (dB)	New System Fig. 24(b) (dB)
Digital Gain	Digital Gain Input	D.R. before Input	230dB (DR6)	140dB (DR6_opt)
	Digital Gain Output	D.R. after Output	70dB (DR7)	70dB (DR7)
	Gain Control Range	-	160dB(DR6-DR7)	70dB(DR6_opt - DR7)

(*1) This estimation is based on virtual model of the Doppler ultrasound system

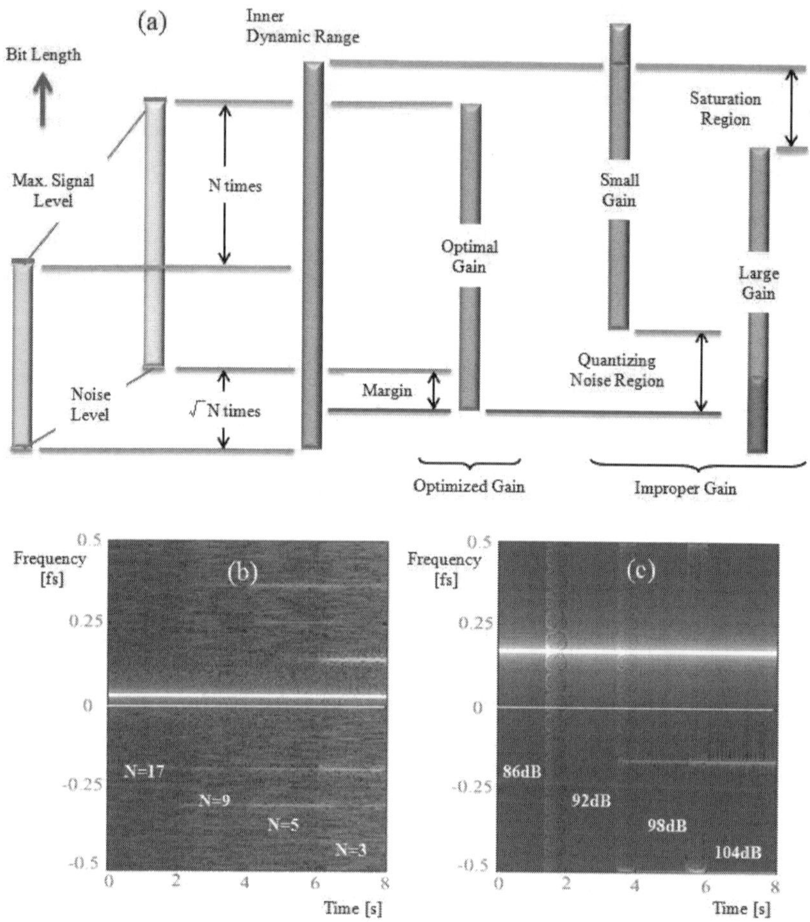

Figure 23. Artifacts caused by inadequate gain control (a) Problems of inadequate gain control (b) Artifacts caused by quantization (c) Artifacts caused by mirror effect.

6.3. Wide dynamic-range design and its optimization

Table 2 depicts the dynamic-range increment and the gain-control range increment of the conventional system and new system based on the signal-processing block diagram shown in Fig. 22. The DBF has the beam-summing effect of *N1* channels. The Mixer/BPF has the band-limitation effect of *f1/f2*. The RG has the integration effect of *N2* taps. The FFT has the integration effect (weighted by the window and operator) of *N3*. The PSD/Pre-Comp. just transfer the dimension (amplitude into power) using the square-root. The MA has the moving average effect of *N4* columns according to the sweep speed of spectrum display. Fig. 24 shows the gain charts of the conventional system and new system based on Table 2. The gain chart of the conventional system which does not take the realization scale of hardware into consideration is shown in Fig. 24(a). This time, I developed the system that can reduce the gain-control range and can also reduce an internal dynamic-range. The automatic gain compensation according to the change-range of system parameters is realized for every sub-block of Doppler signal-processing accompanied by improved S/N ratio. Since the ranges of system parameters are known, the improvement of S/N ratio, the maximum signal level, and the noise level are calculable. An internal dynamic-range scale and the gain adjustment range can be optimally designed for every sub-block. By connecting the partial optimal sub-block in series and uniting it the internal dynamic-range of the system can be reduced so the system size and the total gain control range can both be sharply reduced. The internal S/N ratio is increased by $\sqrt{N}$. Then, supposing an input signal dynamic-range is DRin [dB], the range expansion equivalent to 20*log($\sqrt{N}$) [dB] will occur. Moreover, the internal dynamic-range DRproc [dB] which added more than the margin (20*log($\sqrt{12}$)to quantizing noise is roughly calculable using equation (3).

$$DRproc \geq DRin + 20 \cdot \log\left(\sqrt{N}\right) + 20 \cdot \log\left(\sqrt{12}\right) \tag{3}$$

When digitizing, the Least Significant Bit (LSB) must be rounded not truncated otherwise an error of 1/2*LSB will exist. The RMS value of the quantizing noise is equivalent to 1/2*LSB/ $\sqrt{3}$. So an additional margin of $\sqrt{12}$ for dynamic-range must be maintained so rounding can be performed accurately. Although the internal dynamic-range DRproc is enough only in the automatic gain compensation with system parameters, it is necessary to consider a margin to the original gain adjustment that adjusts for diagnostic target and Doppler sensitivity variation at an internal dynamic-range. The gain chart of the sub-block signal-processing accompanied by range expansion is shown in Fig. 24(b). The gain-control range of the conventional system (corresponds to Fig. 24(a)) is DR6-DR7, and the gain-control range of the new system (corresponds to Fig. 24(b)) is DR6_opt-DR7. In the general Doppler signal-processing DR6 is above 200 dB, and DR7 (the digital gain output) is display luminance range (about 70 dB). So the gain-control range of conventional system should be more than 130 dB. This is very large. On the other hand the DR6_opt of new system is smaller than DR6 about 100 dB. I can reduce not only gain-control range but also the inner dynamic-ranges of sub-modules at the same time [21].

The effect of the automatic gain optimization using the new system was checked in a simulation of RG-integral processing. The spectrum images of the conventional system and the new system when changing RG-width are shown in Fig. 25(a) and Fig. 25(b). The horizontal axis is time and the vertical axis is the frequency normalized by a sampling frequency fs. The range-gate was adjusted from 1mm to 4 mm to 16 mm in 1.8 s intervals. A sinusoidal signal including white noise was used as an input. In the conventional system of Fig. 25(a), the signal level and noise level increase with expanding RG-width. For this reason, the user should reducing gain manually when the RG-width is expanded. In the new system in Fig. 25(b), although the signal level will rise if RG-width is expanded, it turns out that a noise level does not change. As mentioned above, by Doppler automatic gain compensation, the input bit length of each signal-processing block could be made smaller, and also the gain adjustment range could be made small to necessary minimum.

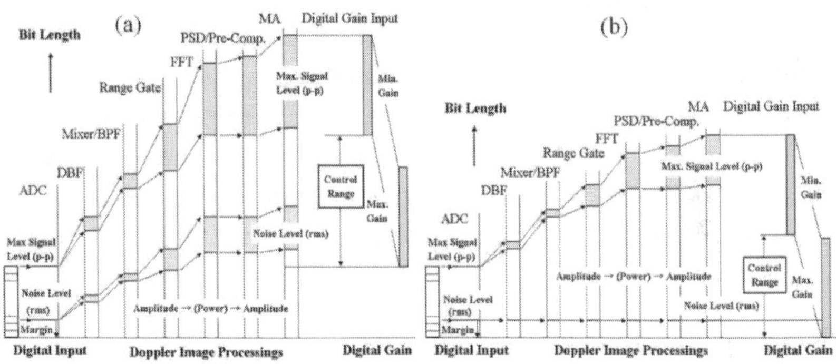

Figure 24. Comparison of gain control systems (a) Gain chart of conventional system (b) Gain chart of new system.

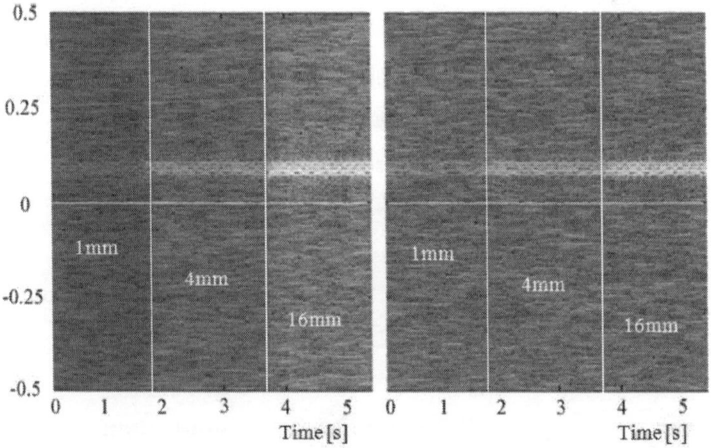

Figure 25. Effect of automatic gain compensation: example of RG integration process. (a) Conventional system (b) New system

7. CONCLUSION

The technical innovations which digital signal-processing brought about and their results were introduced based on some examples of the Doppler ultrasound system architectures. Not only extensive improvement of cost, size, power consumption, and adjustment, but also the improvement of sensitivity and accuracy has been realized by digital technology. Although DSP is most suitable for real-time system at present, the system architecture will be mounted as software when the calculation power of CPU/GPU improves further. In the future we will be able to acquire huge calculation ability easily, and it will be possible to apply it to real-time automatic diagnostic technology etc. besides conventional signal-processing.

REFERENCES

1. T. Baba, 2009 Progress of Doppler ultrasound system architecture and considerations: The problems caused by digital system and their solutions. Society of Signal Processing Applications and Technology of Japan 12 1 2 8

2. T. Baba, 2005 Investigation on direction split technique of Doppler ultrasound: Comparison of six kinds of Doppler audio processing. Society of Signal Processing Applications and Technology of Japan 8 2 14 20

3. T. Baba, 2007 Direction separation in Doppler audio of ultrasound diagnosis equipment: Signal processing for Doppler audio dealiasing. Acoustical Science and Technology 28 3 202 210 DOI: ast.28.202

4. T. Baba, 2005 Research on Doppler ultrasound automatic heartbeat cycle detection: The investigation of heartbeat cycle detection from Doppler waveform using adaptive BPF The Journal of the Acoustical Society of Japan 61 11 629 635

5. T. Baba, 2006 Velocity range tracking in Doppler diagnostic ultrasound systems: Range optimization using Doppler trace wave form histograms. The Journal of the Acoustical Society of Japan 62 4 327 331

6. T. Baba, N. Ohmae, K. Osuka, 2008 The Optimization of Ultrasound System Doppler Velocity Range using Hybrid Control. Transactions of the Society of Instrument and Control Engineers 44 9 760 765

7. T. Ohashi, 2001 Recording of World Heritage on the High Definition Audio-visual Media: Documentation of History and Tradition. ITE Transactions on Media Technology and Applications 55 1 37 46

8. S. Yoshikawa, 2002 Present status of high definition audio. The Journal of the Acoustical Society of Japan 58 4 250 255

9. T. Misawa, 2004 The image sensor for digital cameras. Japanese Journal of Optics 33 9 544 549

10. The Physical Society of Japan. 1978 Physics experiment data processing by a computer, the 4th edition. Tokyo: SAIENSU-SHA Co., Ltd.

11. H. Miyagawa, 1981 Digital Signal Processing The 9th edition. Tokyo: CORONA PUBLISHING Co., LTD.

12. T. Nakajima, 1996 Compact disk reader, the 3rd edition. Tokyo: Ohmsha Ltd.

13. T. Baba, Corp. Toshiba, 2008 Ultrasonic diagnostic equipment. Japanese Patent 4068208; Jan. 18.

14. T. Baba, Corp. Toshiba, 2006 Ultrasonic diagnostic equipment. Japanese Patent 3746113 Feb. 15.

15. JA. Jensen, 1996 Estimation of Blood Velocities Using Ultrasound: A Signal Processing Approach: Cambridge University Press 0-521-46484

16. T. Baba, 2008 Evaluation of Post Wall Filter for Doppler Ultrasound Systems. Acoustical Imaging 29 133 978-1-40208-822-3

17. T. Baba, 2006 Investigation of wall filters in Doppler ultrasound system. Society of Signal Processing Applications and Technology of Japan 9 2 14 19

18. M. Kozak, M. Karaman, 2001 Digital Phased Array Beamforming Using Single-Bit Delta-Sigma Conversion with Non-Uniform Oversampling. IEEE Transactions on UFFC 48 4 922 931 DOI: 10.1109/58.935709

19. S. Engelberg, 2006 Implementing a $\Delta\Sigma$ DAC in Fixed Point Arithmetic. IEEE Signal Processing Magazine: DOI:10.1109/SP-M.2006.248716 66 69

20. T. Baba, Corp. Toshiba, 2005 Ultrasonic diagnostic equipment. Japanese Patent 3663206

21. T. Baba, 2009 Investigation of gain optimization technique in Doppler ultrasound system. Acoustical Science and Technology 30 2 61 71 DOI: ast.30.67

Index